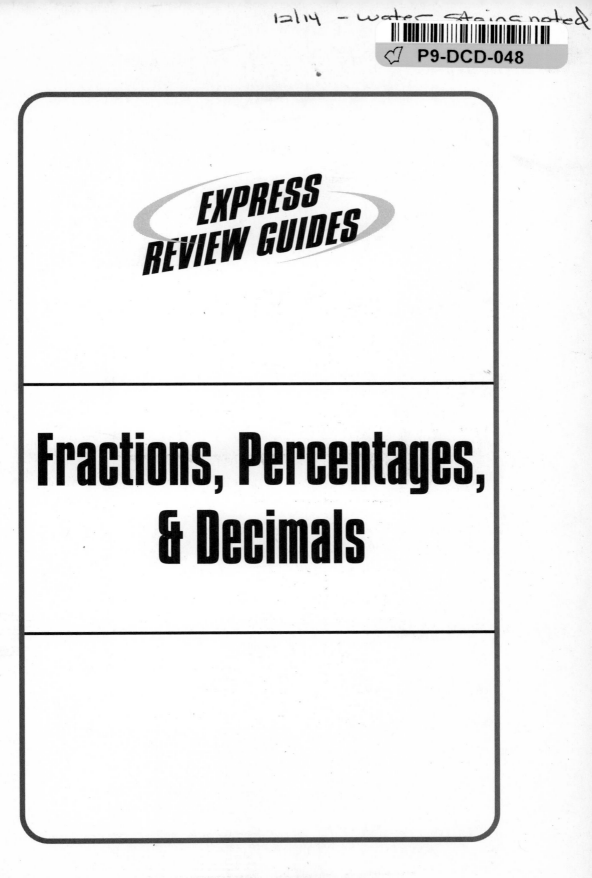

EXPRESS REVIEW GUIDES

Fractions, Percentages, & Decimals

EXPRESS REVIEW GUIDES

Fractions, Percentages, & Decimals

LEARNINGEXPRESS ®

New York

Library of Congress Cataloging-in-Publication Data:

Express review guides : fractions, percentages, & decimals.
        p. cm.
    ISBN: 978-1-57685-621-5 .
1. Fractions. 2. Percentage. 3. Decimal fractions. I. LearningExpress (Organization)
    QA117.E97 2008
    513.2'6—dc22

                                                                    2008000906

Printed in the United States of America

9 8 7 6 5 4 3 2 1

First Edition

ISBN: 978-1-57685-621-5

For more information or to place an order, contact LearningExpress at:
        55 Broadway
        8th Floor
        New York, NY 10006

Or visit us at:
        www.learnatest.com

Contents

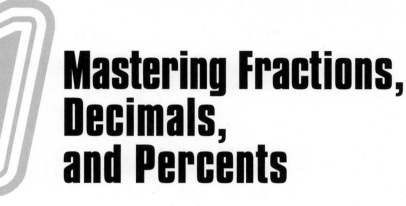

Mastering Fractions, Decimals, and Percents

WHY LEARN ABOUT FRACTIONS, DECIMALS, AND PERCENTS?

Let's say three of eight slices of a pizza have been eaten. What part of the pizza remains? If a sprinter runs a 40-yard dash in more than four seconds but less than five seconds, how can we represent the sprinter's exact time? A student answers 22 of 25 test questions correctly. How can we express the student's grade as a percent on the test?

All of those questions can be answered using fractions, decimals, and percents. We need fractions, decimals, and percents to represent parts of a whole. Many real-life math problems involve whole numbers, but even more involve decimals. Think about a recent purchase you made at a store. Was the exact total at the register a whole number? Coins you receive as change represent decimals—coins represent parts of a whole dollar. The hamburger you ate at a fast-food restaurant probably wasn't a whole pound of beef, but it might have been a quarter-pound. Fractions and food come up pretty often, whether it's pounds of beef, slices of pie, or the weight of a turkey. If you're shopping for a dress, you might be pretty happy if you find one you like, and happier if the tag shows a 30% sale. You'd be even happier if you could figure out exactly how much you were about to save.

It's tough to get through a day without coming across any fractions, decimals, or percents. This book will teach you how to handle every common situation involving parts of numbers (and even a few uncommon ones).

WHY THIS BOOK?

The book begins with fractions, explaining what they are and how to work with them. We build skills slowly, showing the step-by-step procedures to use to solve each type of problem. By reading through the given examples, you will have a strong foundation before beginning the practice problems that follow. Each chapter ends not with just an answer key, but with thorough explanations for every practice problem, so that you are left with more than just a solution—you are left with an understanding of how to solve the problem, and other problems like it. After learning all there is to know about fractions, we move on to decimals and percents before spending a chapter working with all three at once.

The skills covered in each chapter build on the skills taught in the preceding chapters. If you jump right into mixed numbers and find yourself unsure of how to take the reciprocal of a fraction, move back a chapter or two. This book does not assume you have any knowledge of fractions, decimals, or percents. Everything you need to know is found on these pages.

THE GUTS OF THIS GUIDE

Okay, you've obviously cracked open the cover of this book if you're reading these words. But let's take a quick look at what is lurking in the other chapters. This book includes:

- ➡ a 50-question benchmark pretest to help you assess your knowledge of the concepts and skills in this guide
- ➡ brief, focused lessons covering essential basic fraction, percentage, and decimal topics, skills, and applications
- ➡ specific tips and strategies to use as you study
- ➡ a 50-question posttest followed by complete answer explanations to help you assess your progress

As you work through this book, you'll notice that the chapters are sprinkled with all kinds of helpful tips and icons. Look for these icons and the tips they provide. They include:

➡ *Fuel for Thought*: critical information and definitions that can help you learn more about a particular topic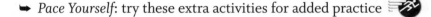

➡ *Practice Lap*: quick practice exercises and activities to let you test your knowledge

➡ *Inside Track*: tips for reducing your study and practice time—without sacrificing accuracy

➡ *Caution!*: pitfalls to be on the lookout for

➡ *Pace Yourself*: try these extra activities for added practice

Ready, Set, Go!

To best use this guide, you should start by taking the pretest. You'll test your math skills and see where you might need to focus your study.

Your performance on the pretest will tell you several important things. First, it will tell you how much you need to study. For example, if you got eight out of ten questions right (not counting lucky guesses!), you might need to brush up only on certain areas of knowledge. But if you got only five out of ten questions right, you will need a thorough review. Second, it can tell you what you know well (that is, which subjects you *don't* need to study). Third, you will determine which subjects you need to study in-depth and which subjects you simply need to refresh your knowledge.

REMEMBER. . .

THE PRETEST IS only practice. If you did not do as well as you anticipated, do not be alarmed and certainly do not despair. The purpose of the quiz is to help you focus your efforts so that you can *improve*. It is important to analyze your results carefully. Look beyond your score, and consider *why* you answered some questions incorrectly. Some questions to ask yourself when you review your wrong answers:

➥ Did you get the question wrong because the material was totally unfamiliar?

➥ Was the material familiar, but you were unable to come up with the right answer? In this case, when you read the right answer, it will often make perfect sense. You might even think, "I knew that!"

➥ Did you answer incorrectly because you read the question carelessly?

Next, look at the questions you got right and review how you came up with the right answer. Not all right answers are created equal.

➥ Did you simply know the right answer?

➥ Did you make an educated guess? An educated guess might indicate that you have some familiarity with the subject, but you probably need at least a quick review.

➥ Did you make a lucky guess? A lucky guess means that you don't know the material and you will need to learn it.

After the pretest, begin the lessons, study the example problems, and try the practice problems. Check your answers as you go along, so if you miss a question, you can study a little more before moving on to the next lesson.

After you've completed all the lessons in the book, try the posttest to see how much you've learned. You'll also be able to see any areas where you may need a little more practice. You can go back to the section that covers that skill for some more review and practice.

THE RIGHT TOOLS FOR THE JOB

BE SURE THAT you have all the supplies you need on hand before you sit down to study. To help make studying more pleasant, select supplies that you enjoy using. Here is a list of supplies that you will probably need:

➡ notebook or legal pad dedicated to studying for your test
➡ pencils
➡ pencil sharpener
➡ highlighter
➡ index or other note cards
➡ paper clips or sticky notepads for marking pages
➡ calendar or personal digital assistant (which you will use to keep track of your study plan)
➡ calculator

As you probably realize, no book can possibly cover all of the skills and concepts you may be faced with. However, this book is not just about building a basic fractions, percentages, and decimals base, but also about building those essential skills that can help you solve unfamiliar and challenging questions. The basic fraction, percentage, and decimal topics and skills in this book have been carefully selected to represent a cross-section of basic skills that can be applied in a more complex setting, as needed.

Pretest

The following exam will test your knowledge of fractions, decimals, and percents and your knowledge of the skills needed to work with these kinds of numbers. The 50 questions are presented in the same order in which the topics they cover are presented in this book. Many of these skills build upon one another, so you may find some of the questions early in the test easier than questions that come later. If you are more comfortable with one area than another, such as percents, you may find those questions easier than others.

The pretest will show you what topics you have mastered, what topics you need to review, and maybe even some topics you need to learn for the first time. If you struggle with or are unable to solve any of these questions, review the answer explanations that follow the test. The explanations also provide the chapter in which the skill being tested is taught.

Save your answers after you have completed the pretest. When you have finished learning the concepts in this book, take the posttest. Compare your pretest to your posttest to see how you've improved.

Answer the questions that follow. Reduce all fractions to their simplest forms and round all decimals to the nearest thousandth.

1. What fraction of the rectangle below is shaded?

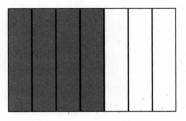

2. Reduce $\frac{16}{24}$ to its simplest form.

3. Find the least common multiple of 10 and 8.

4. Which fraction is greater, $\frac{5}{9}$ or $\frac{7}{12}$?

5. $\frac{6}{11} + \frac{3}{11} =$

6. $\frac{1}{4} + \frac{2}{7} =$

7. $\frac{10}{17} - \frac{6}{17} =$

8. $\frac{3}{4} - \frac{2}{5} =$

9. $\frac{2}{3} \times \frac{9}{13} =$

10. $\frac{5}{6} \div \frac{3}{2} =$

11. Convert $\frac{19}{5}$ to a mixed number.

12. Convert $6\frac{5}{8}$ to an improper fraction.

13. $13\frac{7}{10} + 17\frac{2}{10} =$

14. $23\frac{1}{3} - 6\frac{5}{12} =$

15. $4\frac{5}{6} \times 12\frac{1}{2} =$

16. $3\frac{4}{5} \div 2\frac{1}{7} =$

17. Simplify this complex fraction: $\dfrac{\left(\frac{1}{6}\right)}{\left(\frac{2}{3}\right)}$

18. Write this division sentence as a complex fraction: $\frac{10}{11} \div 8$.

19. Write this ratio as a fraction in simplest form: 25:60.

20. The ratio of cars to trucks in a parking lot is 5:2. If there are 35 cars in the lot, how many trucks are in the lot?

21. In the number 5,436.09817, what number is in the hundredths place?

22. What is the value of the digit 4 in 2,546.577?

23. Write 61.13 in words.

24. Write "one hundred forty and nineteen hundredths" in digits.

25. Round 656.763 to the nearest tenth.

26. Which decimal is greater, 8.452 or 8.461?

27. $2.347 + 12 =$

28. $602.31 - 55.92 =$

29. $4.25 \times 6.1 =$

30. $254.322 \div 9 =$

31. Write 85.417% as a decimal.

32. What is 28% of 234?

33. Write 0.294 as a percent.

34. What percent of 75 is 55?

35. What is 0.32% of 86?

36. What percent of 4 is 19.54?

37. Find the percent increase from 16 to 22.

38. If the value 53 increases by 9%, what is the new value?

39. A value is increased by 11.2% to 72.28. What was the original value?

40. Find the percent decrease from 65.2 to 42.8.

41. If the value 123 decreases by 13%, what is the new value?

42. A value is decreased by 162% to –202.74. What was the original value?

43. Monica buys a dress for $129 and shoes for $44.99. If she pays 8.5% sales tax, what is her total bill?

44. How much interest does a principal of $1,420 gain at a rate of 5.5% for 3 years?

45. Write $42\frac{21}{32}$ as a decimal.

46. Write $\frac{47}{53}$ as a percent.

47. Write 0.951 as a fraction.

48. Write 7,449% as a mixed number.

49. Which is greater, 0.732 or $\frac{8}{11}$?

50. Which of the following numbers are equal?
$\frac{55}{80}$, 0.6875, 6.875%

ANSWERS

1. The rectangle is divided into 7 parts, so the denominator of the fraction will be 7. There are 4 parts of the rectangle that are shaded, so the numerator of the fraction will be 4. Because 4 out of 7 parts are shaded, $\frac{4}{7}$ of the rectangle is shaded. For more on this concept, see Chapter 3.

2. First, list the factors of the numerator and the denominator. The factors of 16 are 1, 2, 4, 8, and 16, and the factors of 24 are 1, 2, 3, 4, 6, 8, 12, and 24. The greatest common factor (the largest number that is a factor of both 16 and 24) is 8. Divide the numerator and denominator of $\frac{16}{24}$ by 8: $\frac{16}{8} = 2$ and $\frac{24}{8} = 3$. $\frac{16}{24}$ reduces to $\frac{2}{3}$. For more on this concept, see Chapter 3.

3. The least common multiple of two numbers is the smallest number that is divisible by both of the two numbers. List the multiples of each number:

   10: 10, 20, 30, **40**, 50, 60, . . .

   8: 8, 16, 24, 32, **40**, 48, 56, . . .

   The smallest number that is a multiple of both 10 and 8 is 40; 40 is the least common multiple of 10 and 8. For more on this concept, see Chapter 3.

4. In order to compare these fractions, we must find common denominators for them, so we must find the least common multiple of 9 and 12. List the multiples of each number:

   9: 9, 18, 27, **36**, 45, 54, 63, . . .

   12: 12, 24, **36**, 48, 60, 72, . . .

   The least common multiple of 9 and 12 is 36. Convert both fractions to a number over 36. Because $\frac{36}{9} = 4$, the new denominator of the fraction $\frac{5}{9}$ is four times larger. Therefore, the new numerator must also be four times larger, so that the value of the fraction does not change: $5 \times 4 = 20$; $\frac{5}{9} = \frac{20}{36}$. Because $\frac{36}{12} = 3$, the new denominator of the fraction $\frac{7}{12}$ is three times larger. Multiply the numerator of the fraction by three: $7 \times 3 = 21$; $\frac{7}{12} = \frac{21}{36}$. Now that we have two fractions with common denominators, we can compare their numerators. Because 21 is greater than 20, $\frac{21}{36} > \frac{20}{36}$ and $\frac{7}{12} > \frac{5}{9}$. For more on this concept, see Chapter 3.

5. These fractions are like fractions, because they have the same denominator. To add two like fractions, add the numerators of the fractions. $6 + 3 = 9$. The denominator of our answer is the same as the denominator of the two fractions that we are adding. Both fractions have a denominator of 11, so the denominator of our answer will be 11; $\frac{6}{11}$ + $\frac{3}{11} = \frac{9}{11}$. For more on this concept, see Chapter 4.

6. Because these fractions are unlike, we must find common denominators for them before adding. First, we find the least common multiple of 4 and 7:

$$4: 4, 8, 12, 16, 20, 24, \mathbf{28}, 32, \ldots$$

$$7: 7, 14, 21, \mathbf{28}, 35, 42, 49, \ldots$$

The least common multiple of 4 and 7 is 28. Convert both fractions to a number over 28. Because $\frac{28}{4} = 7$, the new denominator of the fraction $\frac{1}{4}$ is seven times larger. Therefore, the new numerator must also be seven times larger, so that the value of the fraction does not change: $1 \times 7 = 7. \frac{1}{4} = \frac{7}{28}$. Because $\frac{28}{7} = 4$, the new denominator of the fraction $\frac{2}{7}$ is four times larger. Multiply the numerator of the fraction by four: $2 \times 4 = 8$; $\frac{2}{7} = \frac{8}{28}$. Now that we have like fractions, we can add the numerators: $7 + 8 = 15$. Because the denominator of the fractions we are adding is 28, the denominator of our answer is 28; $\frac{7}{28} + \frac{8}{28} = \frac{15}{28}$. For more on this concept, see Chapter 4.

7. These fractions are like fractions, so we can subtract the numerator of the second fraction from the numerator of the first fraction: $10 - 6 = 4$. Because the denominator of both fractions is 17, the denominator of our answer is 17; $\frac{10}{17} - \frac{6}{17} = \frac{4}{17}$. For more on this concept, see Chapter 4.

8. Because these fractions are unlike, we must find common denominators for them before subtracting. First, we find the least common multiple of 4 and 5:

$$4: 4, 8, 12, 16, \mathbf{20}, 24, \ldots$$

$$5: 5, 10, 15, \mathbf{20}, 25, \ldots$$

The least common multiple of 4 and 5 is 20. Convert both fractions to a number over 20. Because $\frac{20}{4} = 5$, the new denominator of the fraction $\frac{3}{4}$ is five times larger. Therefore, the new numerator must also be five times larger, so that the value of the fraction does not change: $3 \times 5 = 15$: $\frac{3}{4} = \frac{15}{20}$. Because $\frac{20}{5} = 4$, the new denominator of the fraction $\frac{2}{5}$ is four times larger. Multiply the numerator of the fraction by four: $2 \times 4 = 8$. $\frac{2}{5} = \frac{8}{20}$. Now that we have like fractions, we can subtract the second numerator from the first numerator: $15 - 8 = 7$. Because the denominator of the fractions is 20, the denominator of our answer is 20: $\frac{15}{20} - \frac{8}{20} = \frac{7}{20}$. For more on this concept, see Chapter 4.

9. The product of two fractions is equal to the product of the numerators over the product of the denominators. Multiply the numerators: $2 \times 9 = 18$. Multiply the denominators: $3 \times 13 = 39$. $\frac{2}{3} \times \frac{9}{13} = \frac{18}{39}$. We can simplify our answer by dividing the numerator and denominator by their greatest common factor. The greatest common factor of 18 and 39 is 3; $\frac{18}{3} = 6$ and $\frac{39}{3} = 13$, so $\frac{18}{39} = \frac{6}{13}$. For more on this concept, see Chapter 4.

10. To divide a fraction by a fraction, first find the reciprocal of the divisor (the fraction by which you are dividing). In the number sentence $\frac{5}{6} \div \frac{3}{2}$, $\frac{3}{2}$ is the divisor. The reciprocal of a number can be found be switching the numerator with the denominator. The reciprocal of $\frac{3}{2}$ is $\frac{2}{3}$. Now, multiply the dividend ($\frac{5}{6}$) by the reciprocal of the divisor: $\frac{5}{6} \times \frac{2}{3}$. Multiply the numerators: $5 \times 2 = 10$. Multiply the denominators: $6 \times 3 = 18$; $\frac{5}{6} \times \frac{2}{3} = \frac{10}{18}$, so $\frac{5}{6} \div \frac{3}{2} = \frac{10}{18}$. We can simplify our answer by dividing the numerator and denominator by their greatest common factor. The greatest common factor of 10 and 18 is 2: $\frac{10}{2} = 5$ and $\frac{18}{2} = 9$, so $\frac{10}{18} = \frac{5}{9}$. For more on this concept, see Chapter 4.

11. To convert an improper fraction to a mixed number, divide the numerator by the denominator: 19 divided by 5 is 3 with 4 left over. We express the remainder as a fraction. Because the improper fraction has a denominator of 5, our remainder has a denominator of 5; 19 divided by 5 is $3\frac{4}{5}$. For more on this concept, see Chapter 5.

12. To convert a mixed number to an improper fraction, we begin by multiplying the whole number, 6, by the denominator of the fraction, 8: $6 \times 8 = 48$. Next, add to that product the numerator of the fraction: $48 + 5 = 53$. Finally, put that sum over the denominator of the fraction: $6\frac{5}{8} = \frac{53}{8}$. For more on this concept, see Chapter 5.

13. To add two mixed numbers, first, add the whole number parts of each number: $13 + 17 = 30$. Next, add the fractions. Because these fractions

are like, just add the numerators and keep the denominator: $\frac{7}{10} + \frac{2}{10} = \frac{9}{10}$. This sum is a proper fraction, so our answer is $30\frac{9}{10}$. For more on this concept, see Chapter 5.

14. Begin by converting both mixed numbers to improper fractions. Multiply the whole number by the denominator of the fraction, and then add that product to the numerator of the fraction: $23 \times 3 = 69$, $69 + 1 = 70$, so $23\frac{1}{3} = \frac{70}{3}$; $6 \times 12 = 72$, $72 + 5 = 77$, so $6\frac{5}{12} = \frac{77}{12}$. The least common multiple of 3 and 12 is 12, so convert $\frac{70}{3}$ to twelfths. Because $\frac{12}{3} = 4$, multiply the numerator of $\frac{70}{3}$ by four so that the value of the fraction does not change; $70 \times 4 = 280$, $\frac{70}{3} = \frac{280}{12}$. Now we have like fractions: $\frac{280}{12} - \frac{77}{12}$. Subtract the second numerator from the first numerator and keep the denominator. $280 - 77 = 203$, so $\frac{280}{12} - \frac{77}{12} = \frac{203}{12}$. Convert $\frac{203}{12}$ to a mixed number by dividing 203 by 12: $\frac{203}{12} = 16$ with 11 left over. We express the remainder as a fraction. Because the improper fraction has a denominator of 12, our remainder has a denominator of 12; 203 divided by 12 is $16\frac{11}{12}$; $23\frac{1}{3} - 6\frac{5}{12} = 16\frac{11}{12}$. For more on this concept, see Chapter 5.

15. Begin by converting both mixed numbers to improper fractions. Multiply the whole number by the denominator of the fraction, and then add that product to the numerator of the fraction: $4 \times 6 = 24$, $24 + 5 = 29$, so $4\frac{5}{6} = \frac{29}{6}$; $12 \times 2 = 24$, $24 + 1 = 25$, so $12\frac{1}{2} = \frac{25}{2}$. Now we have $\frac{29}{6} \times \frac{25}{2}$. The product of two fractions is equal to the product of the numerators over the product of the denominators. Multiply the numerators: $29 \times 25 = 725$. Multiply the denominators: $6 \times 2 = 12$. $\frac{29}{6} \times \frac{25}{2} = \frac{725}{12}$; $\frac{725}{12} = 60$ with 5 left over. We express the remainder as a fraction. Because the improper fraction has a denominator of 12, our remainder has a denominator of 12; 725 divided by 12 is $60\frac{5}{12}$. $4\frac{5}{6} \times 12\frac{1}{2} = 60\frac{5}{12}$. For more on this concept, see Chapter 5.

16. Begin by converting both mixed numbers to improper fractions. Multiply the whole number by the denominator of the fraction, and then add that product to the numerator of the fraction: $3 \times 5 = 15$, $15 + 4 = 19$, so $3\frac{4}{5} = \frac{19}{5}$; $2 \times 7 = 14$, $14 + 1 = 15$, so $2\frac{1}{7} = \frac{15}{7}$. Now we have $\frac{19}{5} \div \frac{15}{7}$. To divide two fractions, take the reciprocal of the divisor and then multiply. To find the reciprocal of the divisor, $\frac{15}{7}$, switch the numerator and the denominator. The reciprocal of $\frac{15}{7}$ is $\frac{7}{15}$. The problem is now $\frac{19}{5} \times \frac{7}{15}$. Multiply the numerators: $19 \times 7 = 133$. Multiply the denominators:

$5 \times 15 = 75$. $\frac{19}{5} \times \frac{7}{15} = \frac{133}{75}$. $\frac{133}{75} = 1$ with 58 left over. Express the remainder as a fraction. Because the improper fraction has a denominator of 75, our remainder has a denominator of 75; 133 divided by 75 is $1\frac{58}{75}$. $3\frac{4}{5} \div 2\frac{1}{7} = 1\frac{58}{75}$. For more on this concept, see Chapter 5.

17. A complex fraction, like any fraction, represents a division sentence: $\frac{\left(\frac{1}{6}\right)}{\left(\frac{2}{3}\right)}$ $= \frac{1}{6} \div \frac{2}{3}$. To divide two fractions, take the reciprocal of the divisor and then multiply. The reciprocal of $\frac{2}{3}$ is $\frac{3}{2}$. The problem is now $\frac{1}{6} \times \frac{3}{2}$. Divide the 6 in the first fraction and the 3 in the second fraction by 3. The problem becomes $\frac{1}{2} \times \frac{1}{2}$. Multiply the numerators: $1 \times 1 = 1$. Multiply the denominators: $2 \times 2 = 4$. $\frac{1}{2} \times \frac{1}{2} = \frac{1}{4}$. $\frac{1}{6} \div \frac{2}{3} = \frac{1}{4}$. For more on this concept, see Chapter 6.

18. The first fraction, the dividend, becomes the numerator of the fraction and the whole number, the divisor, becomes the denominator of the fraction. $\frac{10}{11} \div 8 = \frac{\left(\frac{10}{11}\right)}{8}$. For more on this concept, see Chapter 6.

19. First, find the greatest common factor of 25 and 60. The factors of 25 are 1, **5**, and 25. The factors of 60 are 1, 2, 3, 4, **5**, 6, 10, 12, 15, 20, 30, and 60. The greatest common factor is 5, so divide both 25 and 60 by 5; $\frac{25}{5} = 5$ and $\frac{60}{5} = 12$. The ratio 25:60 can be reduced to 5:12. When written as a fraction, the first number in the ratio is the numerator and the second number is the denominator: $5:12 = \frac{5}{12}$. For more on this concept, see Chapter 6.

20. Write the ratio as a fraction: $5:2 = \frac{5}{2}$. Set up a proportion. If there are 5 cars for every 2 trucks, then there are 35 cars for x number of trucks: $\frac{5}{2} = \frac{35}{x}$. Cross multiply: $(5)(x) = (2)(35)$, $5x = 70$. Divide both sides of the equation by 5: $\frac{5x}{5} = \frac{70}{5}$, $x = 14$. If there are 35 cars in the parking lot, then there are 14 trucks in the lot. For more on this concept, see Chapter 6.

21. The hundredths place is the second place to the right of the decimal point. In the number 5,436.09817, the 9 is the digit in the second place to the right of the decimal point. For more on this concept, see Chapter 7.

22. The 4 is in the tens place, so it has a value of $4 \times 10 = 40$. For more on this concept, see Chapter 7.

23. Sixty-one and thirteen hundredths. For more on this concept, see Chapter 7.

24. 140.19. For more on this concept, see Chapter 7.

25. The digit to the immediate right of the tenths place is the digit in the hundredths place, 6. Because 6 is greater than 5, we round up. The tenths digit increases by 1 to 8, and the digits to the right become zero; 656.763 to the nearest tenth is 656.800, or 656.8. For more on this concept, see Chapter 7.

26. To compare two decimals, line up the decimal points and compare corresponding digits from left to right:

8.452

8.461

Both numbers have an 8 in the ones place and a 4 in the tenths place, so we move to the hundredths place. The first number has a 5 in the hundredths place and the second number has a 6 in the hundredths place. Because 6 is greater than 5, 8.461 is greater than 8.452. For more on this concept, see Chapter 7.

27. To add a decimal and a whole number, write the problem vertically (be sure to line up the decimal points), put a decimal on the end of the whole number, and place trailing zeros next to it:

2.347

+ 12.000

Add column by column and carry the decimal point down into your answer.

2.347

+ 12.000

14.347

For more on this concept, see Chapter 8.

28. To subtract a decimal from a decimal, write the problem vertically, lining up the decimal points, and then subtract column by column:

602.31

− 55.92

546.39

For more on this concept, see Chapter 8.

29. 4.25 has two digits to the right of the decimal point and 6.1 has one digit to the right of the decimal point, so our answer will have three digits to the right of the decimal point:

$$\begin{array}{r} 4.25 \\ \times\, 6.1 \\ \hline 425 \\ +\,2550 \\ \hline 25.925 \end{array}$$

For more on this concept, see Chapter 8.

30.

$$\begin{array}{r} 28.258 \\ 9{\overline{\smash{\big)}\,254.322}} \\ -18 \\ \hline 74 \\ -72 \\ \hline 23 \\ -18 \\ \hline 52 \\ -45 \\ \hline 72 \\ -72 \\ \hline 0 \end{array}$$

For more on this concept, see Chapter 8.

31. To write a percent as a decimal, move the decimal point two places to the left: 85.417% = 0.85417. For more on this concept, see Chapter 9.

32. First, write 28% as a decimal: 28% = 0.28. Then, multiply the decimal by the number: $0.28 \times 234 = 65.52$. For more on this concept, see Chapter 9.

33. To write a decimal as a percent, move the decimal point two places to the right. 0.294 = 29.4%. For more on this concept, see Chapter 9.

34. To find what percent of 75 is 55, divide 55 by 75: $\frac{55}{75}$ = 0.73333. . ., or 0.733 to the nearest thousandth. To write 0.733 as a percent, move the decimal point two places to the right: 0.733 = 73.3%. For more on this concept, see Chapter 9.

35. We follow the same steps we used when working with percents that are greater than 1. Write the percent as a decimal and multiply. Move the decimal point two places to the left: 0.32% = 0.0032. 0.0032 × 86 = 0.2752, or 0.275 to the nearest thousandth. For more on this concept, see Chapter 9.

36. Divide 19.54 by 4: 19.54 ÷ 4 = 4.885. Write the decimal as a percent by moving the decimal point two places to the right: 4.885 = 488.5%; 19.54 is 488.5% of 4. For more on this concept, see Chapter 9.

37. The original value is 16 and the new value is 22. Subtract the original value from the new value: 22 − 16 = 6. Next, divide the difference by the original value: 6 ÷ 16 = 0.375. Now, write the decimal as a percent by moving the decimal point two places to the right: 0.375 = 37.5%. For more on this concept, see Chapter 10.

38. The percent increase is 9%, or 0.09. Add 1 and multiply by the original value: 53 × 1.09 = 57.77. For more on this concept, see Chapter 10.

39. Write the percent increase as a decimal: 11.2% = 0.112. Add 1 to the percent increase and divide the new value by that sum: 72.28 ÷ 1.112 = 65. For more on this concept, see Chapter 10.

40. Subtract the new value from the original value and divide by the original value: 65.2 − 42.8 = 22.4, 22.4 ÷ 65.2 = 0.343558. . ., or 0.344 to the nearest thousandth. 0.344 = 34.4%. For more on this concept, see Chapter 10.

41. The original value is 123 and the percent decrease is 13%, or 0.13. **(new value) = (1 − percent decrease) × (original value).** Substitute the values into the formula: (1 − 0.13) × 123 = 0.87 × 123 = 107.01. For more on this concept, see Chapter 10.

42. Write the percent decrease as a decimal: 1.62. **(original value) = (new value) ÷ (1 − percent decrease).** Subtract the percent decrease from 1 and divide the new value by that difference: −202.74 ÷ −0.62 = 327. For more on this concept, see Chapter 10.

43. To solve a sales tax problem, begin by adding the costs of each item purchased: $129 + $44.99 = $173.99. Then, convert the sales tax percent to a decimal, and add 1 to it: 8.5% = 0.085, 0.085 + 1 = 1.085. Finally, multiply that decimal by the cost of the items to find the total cost: 1.085 × $173.99 = $188.77915, or $188.78 to the nearest cent. For more on this concept, see Chapter 10.

44. The formula for finding interest is $I = prt$, where I is interest, p is principal, r is the rate, and t is the time in years. Write the rate as a decimal and substitute the values into the formula. 5.5% = 0.055. $I = $1,420 × 0.055 × 3 = $234.30. For more on this concept, see Chapter 10.

45. To write a mixed number as a decimal, begin by writing the whole number part, 42, to the left of the decimal point. Then, write the fraction part as a decimal by dividing the numerator by the denominator. $21 \div 32 = 0.65625$, or 0.656 to the nearest thousandth, so $42\frac{21}{32} = 42.656$, to the nearest thousandth. For more on this concept, see Chapter 11.

46. To write a fraction as a percent, first write the fraction as a decimal by dividing the numerator by the denominator $47 \div 53 = 0.886792\ldots$, or 0.887 to the nearest thousandth. Now, write the decimal as a percent by moving the decimal point two places to the right and adding the percent symbol. 0.887 = 88.7%. For more on this concept, see Chapter 11.

47. To write a decimal as a fraction, first read the name of the decimal out loud: 0.951 is "nine hundred fifty-one thousandths." The first part of the name, *nine hundred fifty-one* (the digits that appear to the right of the decimal point), is the numerator of the fraction. The second part of the name, *thousandths*, is the denominator of the fraction. $0.951 = \frac{951}{1,000}$. For more on this concept, see Chapter 11.

48. To write a percent as a mixed number, first write the percent as a decimal by moving the decimal point two places to the left. 7,449% = 74.49. Now, write the decimal as a mixed number. Read the name of the decimal aloud. 74.49 is "seventy-four and forty-nine hundredths." The whole number part of the mixed number is the part that comes before "and" in the decimal name. The fraction part of the mixed number is forty-nine hundredths. The first part of that name, *forty-nine*, is the numerator of the fraction, and the second part of that name, *hundredths*, is the denominator of the fraction. $74.49 = 74\frac{49}{100}$. For more on this concept, see Chapter 11.

49. Write the fraction $\frac{8}{11}$ as a decimal so that it can be compared to 0.732. Because the decimal 0.732 has a digit in the thousandths place, we must round $\frac{8}{11}$ to the nearest ten thousandth: $8 \div 11 = 0.72727272. . .$, or 0.7273 to the nearest ten thousandth. Add a trailing zero to 0.732, line up the decimal points, and compare corresponding digits of the two numbers from left to right:

$$0.7320$$
$$0.7273$$

Both numbers have a 0 in the ones place and a 7 in the tenths place, so we move to the hundredths place. The first number, 0.7320, has a 3 in the hundredths place, and the second number, 0.7273, has a 2 in the hundredths place. Because $3 > 2$, $0.7320 > 0.7273$, and $0.732 > \frac{8}{11}$. For more on this concept, see Chapter 11.

50. Convert the fraction and the percent to decimals. The fraction can be converted to a decimal by dividing. $\frac{55}{80} = 55 \div 80 = 0.6875$, which is equal to the given decimal, 0.6875; 6.875% can be written as a decimal by removing the percent sign and shifting the decimal point two places to the left; 6.875% = 0.06875, which is not equal to the given fraction or the given decimal. For more on this concept, see Chapter 11.

What's a Fraction?

WHAT'S AROUND THE BEND

- ➡ Defining a Fraction
- ➡ Parts of a Fraction
- ➡ Writing Fractions
- ➡ Using Fractions to Represent Real Situations
- ➡ Proper and Improper Fractions
- ➡ Like and Unlike Fractions
- ➡ Comparing Fractions
- ➡ Equivalent Fractions
- ➡ Simplifying Fractions

DEFINING A FRACTION

What is a fraction, and why do we need them? A **fraction** is a type of number. We use fractions to represent a part of a whole. Let's say a whole pie is made up of eight slices. If you eat two slices, what part of the whole pie have you eaten? Fractions help us answer that question. You have eaten two out of eight slices, or two-eighths of the pie.

The following circle represents our pie. You can see that the circle is divided into eight equal slices. Two of those slices are shaded. The shaded area is two-eighths of the pie, which can be written as $\frac{2}{8}$.

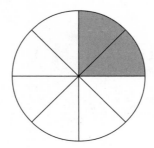

Every fraction is made up of two numbers with a horizontal line between them. The top number is called the **numerator**. The bottom number is called the **denominator**. In the fraction $\frac{2}{8}$, the numerator is 2 and the denominator is 8. The numerator of a fraction can be any number, including zero, but the denominator of a fraction can never be zero. A fraction with a denominator of zero is undefined.

FUEL FOR THOUGHT

A FRACTION REPRESENTS a part of a whole. A fraction itself is a division statement. The top number of the fraction, the **numerator**, is divided by the bottom number of the fraction, the **denominator**.

The denominator is the number of parts that make the whole. Look again at our pie. There are 8 parts, or slices, that make up the whole. If the pie had only 4 slices, then the denominator of our fraction would be 4. The numerator is the number of parts that are shaded. Because our pie has 2 slices shaded, the numerator of our fraction is 2. The numerator always tells us how many parts of the whole we have.

Look at the following fraction. What fraction of this circle is shaded?

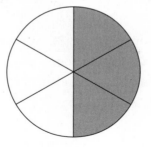

There are 3 parts of the circle that are shaded. The numerator of our fraction will be 3. This circle is divided into only 6 equal parts, so the denominator of our fraction will be 6. Because there are 3 parts that are shaded out of 6 total parts, we can say that $\frac{3}{6}$ of the circle is shaded.

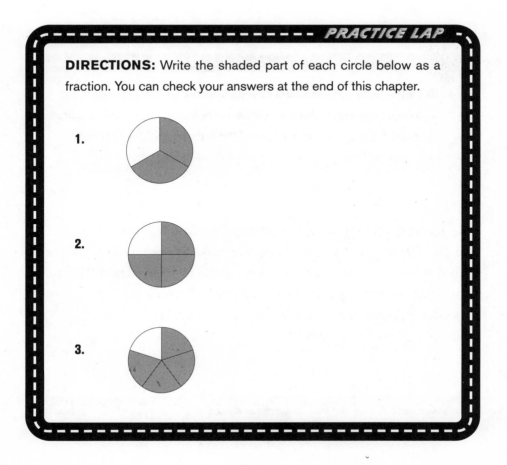

PRACTICE LAP

DIRECTIONS: Write the shaded part of each circle below as a fraction. You can check your answers at the end of this chapter.

1.

2.

3.

We just saw how to turn pictures (shaded circles) into fractions. Now let's go in the opposite direction: Let's draw pictures to represent fractions.

When you look at a fraction, read it as "the numerator out of the denominator." The fraction $\frac{2}{4}$ is "2 out of 4." To represent the fraction $\frac{2}{4}$ with a circle, we need to show 2 out of 4 parts shaded. First, draw a circle divided into 4 equal parts. Then, shade 2 of those parts:

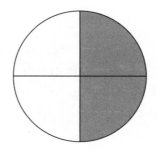

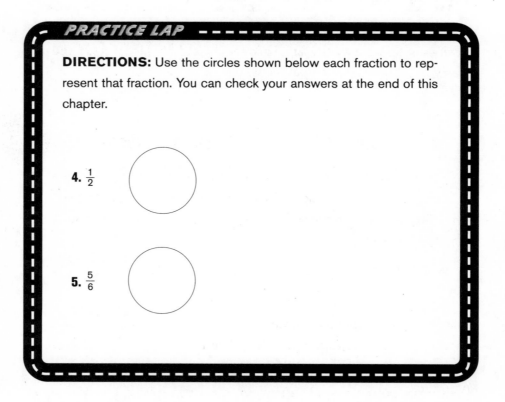

PRACTICE LAP

DIRECTIONS: Use the circles shown below each fraction to represent that fraction. You can check your answers at the end of this chapter.

4. $\frac{1}{2}$

5. $\frac{5}{6}$

PACE YOURSELF

FRACTIONS HELP US describe everyday situations. If you have 15 math problems for homework, and you've completed 5 of them, what fraction of your math homework is complete? The answer: 5 out of 15, or $\frac{5}{15}$.

If $\frac{12}{25}$ of your class has brown hair, and there are 25 students in your class, how many have brown hair? Because the fraction $\frac{12}{25}$ means "12 out of 25," there are 12 students in your class with brown hair.

Find three other real-life scenarios that could be described with fractions.

TYPES OF FRACTIONS

So far, every fraction we have looked at has been a **proper fraction**. A proper fraction has a value that is between –1 and 1. In other words, the part of the whole is always less than the whole.

INSIDE TRACK

HOW CAN YOU determine if a fraction is proper? First, ignore any positive or negative signs. Then, compare the numerator to the denominator. If the numerator is less than the denominator, the fraction is proper.

What do you call a fraction whose numerator is greater than or equal to its denominator? An **improper fraction**. The fractions $\frac{12}{3}$, $\frac{8}{7}$, and $\frac{2}{2}$ are all improper fractions. Later, we'll see how to turn improper fractions into mixed numbers.

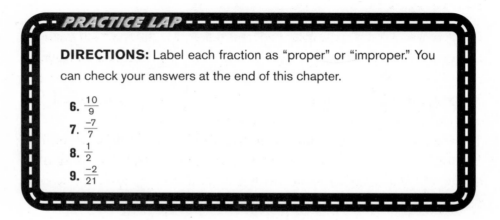

FUEL FOR THOUGHT

A PROPER FRACTION has a value less than 1 and greater than −1. For instance, $\frac{2}{3}$ is a proper fraction, as is $\frac{-2}{3}$. An **improper** fraction has a value greater than or equal to 1, or less than or equal to −1. $\frac{3}{2}$ and $\frac{-3}{2}$ are both improper fractions. Any number over itself is equal to 1. $\frac{12}{12}$, $\frac{1,000}{1,000}$, and $\frac{-8}{-8}$ are all equal to 1.

PRACTICE LAP

DIRECTIONS: Label each fraction as "proper" or "improper." You can check your answers at the end of this chapter.

6. $\frac{10}{9}$

7. $\frac{-7}{7}$

8. $\frac{1}{2}$

9. $\frac{-2}{21}$

Every fraction can be described as either proper or improper. We also have a way of describing pairs of fractions. If two fractions have the same denominator, then they are **like fractions**. If two fractions have different denominators, then they are **unlike fractions**.

The fractions $\frac{1}{4}$ and $\frac{3}{4}$ are like fractions. Both have a denominator of 4. Like fractions are easy to compare: The fraction with the greater numerator is the greater fraction. The fractions $\frac{1}{4}$ and $\frac{2}{3}$ are unlike fractions, because they do not have the same denominator. Unlike fractions are harder to compare. In fact, you should always convert unlike fractions to like fractions before comparing.

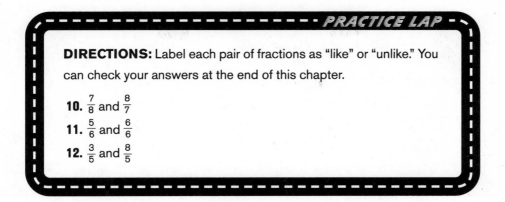

DIRECTIONS: Label each pair of fractions as "like" or "unlike." You can check your answers at the end of this chapter.

10. $\frac{7}{8}$ and $\frac{8}{7}$

11. $\frac{5}{6}$ and $\frac{6}{6}$

12. $\frac{3}{5}$ and $\frac{8}{5}$

COMPARING FRACTIONS

Often, you will want to compare two fractions to determine which fraction is larger. Let's begin by comparing like fractions. Which fraction is bigger, $\frac{2}{5}$ or $\frac{4}{5}$? The fractions are like, so all we need to do is compare the numerators. Because 4 is greater than 2, $\frac{4}{5}$ is greater than $\frac{2}{5}$. What about $\frac{7}{11}$ and $\frac{6}{11}$? Because 7 is greater than 6, $\frac{7}{11}$ is greater than $\frac{6}{11}$.

Comparing unlike fractions is a bit trickier. Think of comparing unlike fractions as comparing two different units of measure. Which length is longer, 28 centimeters or 11 inches? It is very difficult to say—but it would be much easier to figure out if both lengths were given in centimeters, or if both lengths were given in inches. That's why we always turn unlike fractions into like fractions before comparing them.

How do we turn unlike fractions into like fractions? We find a **common denominator**, and rewrite both fractions with that new denominator. The best way to find a common denominator for two unlike fractions is to find the **least common multiple** of those two denominators.

> ### FUEL FOR THOUGHT
>
> **A COMMON DENOMINATOR** for two fractions is a number that is a multiple of each of the denominators of those fractions. For instance, 15 is a common denominator for the fractions $\frac{2}{5}$ and $\frac{1}{3}$ because 15 is a multiple of 5 and a multiple of 3. 15 isn't the only multiple that 5 and 3 have in common—they also have 30, 45, 60, and many others in common—but 15 is the **least common multiple**. Of all the multiples that 3 and 5 have in common, 15 is the smallest.

That may sound hard, but it's as simple as remembering your times tables. Take the unlike fractions $\frac{2}{4}$ and $\frac{4}{6}$. We need to find a common denominator for these fractions before we can compare them. To find the least common multiple of 4 and 6, think of the 4 times table and the 6 times table:

   4 times table: 4, 8, **12**, 16, 20, **24**, . . .
   6 times table: 6, **12**, 18, **24**, 30, 36, . . .

By listing the multiples of 4 and 6, we can find the numbers that are multiples of both 4 and 6. Notice the numbers that appear in both times tables: 12 and 24. These numbers are common multiples of 4 and 6. 12 is the smallest multiple that is common to 4 and 6. We say that 12 is the "least common multiple" of 4 and 6.

Now that we've found the least common multiple, we can convert both fractions into twelfths—fractions with denominators of 12. To do this, we must change both the numerator and the denominator of each fraction.

CAUTION!

THERE ARE INFINITELY many ways to write the value of a single fraction. When converting a fraction to a new fraction with a different denominator, be sure that the value of the fraction does not change. Whatever you do to the denominator, you must also do to the numerator. If you create a new fraction with a denominator that is 10 times greater than the denominator of your original fraction, then you must multiply the numerator of your original fraction by 10 to find the numerator of your new fraction.

We know that the denominators of our new fractions will be 12. But what will the numerators be? Let's start with $\frac{2}{4}$. The denominator of this fraction is 4. The new denominator will be 12. That means that the new denominator is 3 times larger than the old denominator. How did we figure that out? By dividing the new denominator, 12, by the old denominator, 4. Because the new denominator is 3 times larger than the old denominator, the new numerator must be 3 times larger than the old numerator. Because $2 \times 3 = 6$, the numerator of our new fraction is 6. We had to change the numerator in the same way we changed the denominator: $\frac{2}{4} = \frac{6}{12}$. These fractions may look different, but they have the exact same value. In fact, we could have multiplied the numerator and the denominator by 6, creating the fraction $\frac{12}{24}$, and that would have been equal to $\frac{2}{4}$ (and $\frac{6}{12}$) too. As long as you multiply the numerator by the same number by which you multiply the denominator, the value of the fraction will not change.

FUEL FOR THOUGHT

TWO FRACTIONS THAT have the same value are called **equivalent fractions**. The fractions $\frac{1}{3}$, $\frac{2}{6}$, and $\frac{3}{9}$ are all equivalent, or equal, to each other.

TWO FRACTIONS CAN be equal, or equivalent, without being like fractions. For instance, the fractions $\frac{1}{2}$ and $\frac{3}{6}$ are equivalent, but unlike. Also, two fractions can be like, but not equivalent. The fractions $\frac{1}{3}$ and $\frac{2}{3}$ are like, but unequal. In fact, like fractions are equal only if their numerators are equal.

Now let's convert $\frac{4}{6}$ to a fraction with a denominator of 12. Follow the same steps. Divide the new denominator, 12, by the old denominator, 6: $\frac{12}{6} = 2$. The new denominator is 2 times bigger than the old denominator. Therefore, the new numerator must be 2 times bigger than the old numerator: $4 \times 2 = 8$. The numerator of our new fraction is 8. $\frac{4}{6} = \frac{8}{12}$.

Now that we have rewritten our two unlike fractions, $\frac{2}{4}$ and $\frac{4}{6}$, as the like fractions $\frac{6}{12}$ and $\frac{8}{12}$, we are ready to compare them. Because 8 is greater than 6, $\frac{8}{12}$ is greater than $\frac{6}{12}$, and $\frac{4}{6}$ is greater than $\frac{2}{4}$.

Let's look at another example. Which is bigger, $\frac{3}{5}$ or $\frac{5}{10}$? We follow the same steps as in the previous example. First, we find the least common multiple of 5 and 10:

5 times table: 5, **10**, 15, **20**, 25, **30**, . . .
10 times table: **10, 20, 30,** 40, 50, 60, . . .

Every multiple of 10, including 10 itself, is a multiple of 5. Because 10 is the least common multiple of 5 and 10, we don't have to change the fraction $\frac{5}{10}$ at all. We just need to convert $\frac{3}{5}$ to tenths. Divide the new denominator, 10, by 5: $\frac{10}{5} = 2$. Because the new denominator is 2 times bigger than the old denominator, the new numerator must be 2 times bigger than the old numerator: $3 \times 2 = 6$. $\frac{3}{5} = \frac{6}{10}$. Now we are ready to compare. Because 6 is greater than 5, $\frac{6}{10}$ is greater than $\frac{5}{10}$, so $\frac{3}{5}$ is greater than $\frac{5}{10}$.

INSIDE TRACK

HERE ARE SOME tips for comparing fractions:

1. If two fractions have the same numerator, but different denominators, the fraction with the *smaller* denominator is the *bigger* fraction. For instance, $\frac{1}{2} > \frac{1}{3}$. Check by finding common denominators: $\frac{1}{2} = \frac{3}{6}$, and $\frac{1}{3} = \frac{2}{6}$. $\frac{1}{2}$ is greater than $\frac{1}{3}$. $\frac{1}{2}$ is also greater than $\frac{1}{4}$, $\frac{1}{5}$, and $\frac{1}{6}$.

2. If the denominator of one or both of the fractions is a prime number and the other denominator is not a multiple of that prime number, the least common denominator will be the product of the two denominators. For instance, if you are comparing $\frac{1}{5}$ and $\frac{3}{8}$, the least common denominator will be 40 (5 × 8), because 5 is a prime number and 8 is not a multiple of 5. Check by listing the multiples of each:

   5 times table: 5, 10, 15, 20, 25, 30, 35, **40**, . . .
   8 times table: 8, 16, 32, **40**, . . .

3. If two fractions are positive and one of them is improper while the other is proper, the improper fraction is greater—you don't even have to find common denominators. All positive improper fractions are equal to 1 or more, while all positive proper fractions are equal to less than 1.

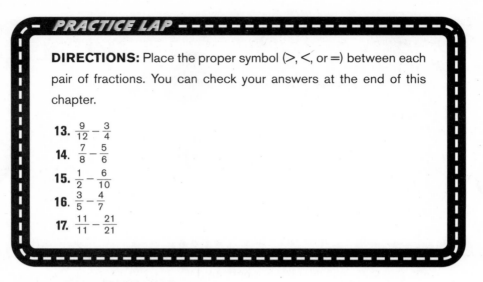

- PRACTICE LAP -

DIRECTIONS: Place the proper symbol (>, <, or =) between each pair of fractions. You can check your answers at the end of this chapter.

13. $\frac{9}{12} - \frac{3}{4}$

14. $\frac{7}{8} - \frac{5}{6}$

15. $\frac{1}{2} - \frac{6}{10}$

16. $\frac{3}{5} - \frac{4}{7}$

17. $\frac{11}{11} - \frac{21}{21}$

SIMPLIFYING FRACTIONS

We've seen how to take fractions and convert them to new fractions with larger denominators. Now, let's go in the other direction. We can make the numbers in the fractions smaller, or simpler, by reducing them without changing the value of the fraction.

We can reduce a fraction by dividing its numerator and denominator by the same number. Remember, whatever we do to the numerator, we must also do to the denominator. In order to reduce a fraction to its simplest form, we must divide the numerator and the denominator by the largest number that is a factor of both the numerator and the denominator. This number is called the **greatest common factor**.

FUEL FOR THOUGHT

THE GREATEST COMMON factor of two numbers is the largest number that divides evenly into both numbers. For instance, the greatest common factor of 12 and 18 is 6. Both 12 and 18 can be divided evenly by 6. Other numbers (1, 2, and 3) are also factors of both 12 and 18, but 6 is the greatest common factor.

Example

Let's simplify the fraction $\frac{8}{12}$. We begin by listing the factors of each number:

Factors of 8: **1, 2, 4,** 8

Factors of 12: **1, 2,** 3, **4,** 6, 12

The factors that 8 and 12 have in common are 1, 2, and 4. There-fore, 4 is the greatest common factor. We can reduce $\frac{8}{12}$ to its sim-plest form if we divide its numerator and denominator by 4: $\frac{8}{4} = 2$ and $\frac{12}{4} = 3$. Therefore, $\frac{8}{12} = \frac{2}{3}$. $\frac{2}{3}$ is $\frac{8}{12}$ in its simplest form.

Example

Now let's simplify $\frac{12}{36}$. Again, begin by listing the factors of each number:

Factors of 12: **1, 2, 3, 4, 6, 12**

Factors of 36: **1, 2, 3, 4, 6,** 9, **12,** 18, 36

12 and 36 have many common factors, but 12 is the greatest com-mon factor. We can reduce $\frac{12}{36}$ to its simplest form if we divide its numerator and denominator by 12: $\frac{12}{12} = 1$ and $\frac{36}{12} = 3$. Therefore, $\frac{12}{36} = \frac{1}{3}$. $\frac{1}{3}$ is $\frac{12}{36}$ in its simplest form.

If a proper fraction has a prime number in its denominator, then it is already in simplest form. However, a proper fraction could have a composite num-ber in its denominator, and it may also be in simplest form. For instance, the fraction $\frac{5}{8}$ is in simplest form even though 8 is not a prime number, but the fraction $\frac{20}{41}$ must be in simplest form, because 41 is a prime number.

CAUTION!

FINDING THE GREATEST common factor of two numbers is not the same as finding the least common multiple of two numbers. When you are simplifying a fraction, you are looking for the greatest common factor between a numerator and denominator of a single fraction. When you are finding common denominators for a pair of fractions, you are looking for the least common multiple of the denominators of the fractions.

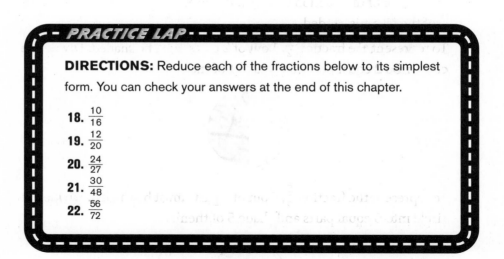

PRACTICE LAP

DIRECTIONS: Reduce each of the fractions below to its simplest form. You can check your answers at the end of this chapter.

18. $\frac{10}{16}$

19. $\frac{12}{20}$

20. $\frac{24}{27}$

21. $\frac{30}{48}$

22. $\frac{56}{72}$

We've seen how to write fractions and how to express real-life situations as fractions. We've also learned to classify fractions as proper or improper and how to classify pairs of fractions as like or unlike. By converting unlike fractions to like fractions, we saw how to compare them. In the next chapter, we'll use these skills to help us add and subtract fractions. Many tests ask for answers to be put in simplest form. Now that we know how to reduce fractions, we can always write our answers in simplest form. We're ready to start REALLY working with fractions.

ANSWERS

1. This circle is divided into 3 equal parts, so the denominator of the fraction will be 3. There are 2 parts of the circle that are shaded, so the numerator of the fraction will be 2. Because 2 out of 3 parts are shaded, $\frac{2}{3}$ of the circle is shaded.

2. This circle is divided into 4 equal parts, so the denominator of the fraction will be 4. There are 3 parts of the circle that are shaded, so the numerator of the fraction will be 3. Because 3 out of 4 parts are shaded, $\frac{3}{4}$ of the circle is shaded.

3. This circle is divided into 5 equal parts, so the denominator of the fraction will be 5. There are 4 parts of the circle that are shaded, so the numerator of the fraction will be 4. Because 4 out of 5 parts are shaded, $\frac{4}{5}$ of the circle is shaded.

4. To represent the fraction $\frac{1}{2}$, 1 out of 2 parts must be shaded. Divide the circle into 2 equal parts and shade 1 of them:

5. To represent the fraction $\frac{5}{6}$, 5 out of 6 parts must be shaded. Divide the circle into 6 equal parts and shade 5 of them:

6. The numerator of this fraction is greater than its denominator. Because the fraction has a value that is greater than 1, it is improper.

7. The numerator of this fraction is the same as its denominator. Because the fraction has a value of –1, it is improper.

8. The numerator of this fraction is less than its denominator. Because the fraction has a value that is less than 1 and greater than –1, it is proper.

9. The numerator of this fraction is less than its denominator. Because the fraction has a value that is less than 1 and greater than –1, it is proper.

10. The denominators of these fractions are different, so these are unlike fractions.

11. The denominators of these fractions are the same, so these are like fractions.

12. Even though the second fraction is improper, the denominators of these fractions are the same, so these fractions are like.

13. The least common denominator for $\frac{9}{12}$ and $\frac{3}{4}$ is 12. $\frac{3}{4} = \frac{9}{12}$ (both the numerator and denominator are 3 times bigger).

14. The least common denominator for $\frac{7}{8}$ and $\frac{5}{6}$ is 24. $\frac{7}{8} = \frac{21}{24}$ (both the numerator and denominator are 3 times bigger) and $\frac{5}{6} = \frac{20}{24}$ (both the numerator and denominator are 4 times bigger). Because 21 is greater than 20, $\frac{21}{24} > \frac{20}{24}$, so $\frac{7}{8} > \frac{5}{6}$.

15. The least common denominator for $\frac{1}{2}$ and $\frac{6}{10}$ is 10. $\frac{1}{2} = \frac{5}{10}$ (both the numerator and denominator are 5 times bigger). Because 5 is less than 6, $\frac{5}{10} < \frac{6}{10}$, so $\frac{1}{2} < \frac{6}{10}$.

16. The least common denominator for $\frac{3}{5}$ and $\frac{4}{7}$ is 35. $\frac{3}{5} = \frac{21}{35}$ (both the numerator and denominator are 7 times bigger) and $\frac{4}{7} = \frac{20}{35}$ (both the numerator and denominator are 5 times bigger). Because 21 is greater than 20, $\frac{21}{35} > \frac{20}{35}$, so $\frac{3}{5} > \frac{4}{7}$.

17. Any number over itself is equal to 1, so both fractions, $\frac{11}{11}$ and $\frac{21}{21}$, are equal to 1, so $\frac{11}{11} = \frac{21}{21}$.

18. First, list the factors of the numerator and the denominator. The factors of 10 are 1, **2**, 5, and 10, and the factors of 16 are 1, **2**, 4, 8, and 16. The greatest common factor is 2: $\frac{10}{2} = 5$ and $\frac{16}{2} = 8$; $\frac{10}{16}$ reduces to $\frac{5}{8}$.

19. List the factors of the numerator and the denominator. The factors of 12 are 1, 2, 3, **4**, 6, and 12, and the factors of 20 are 1, 2, **4**, 5, and 20. The greatest common factor is 4. $\frac{12}{4} = 3$ and $\frac{20}{4} = 5$. $\frac{12}{20}$ reduces to $\frac{3}{5}$.

20. List the factors of the numerator and the denominator. The factors of 24 are 1, 2, **3**, 4, 6, 8, 12, and 24, and the factors of 27 are 1, **3**, 9, and 27. The greatest common factor is 3. $\frac{24}{3} = 8$ and $\frac{27}{3} = 9$. $\frac{24}{27}$ reduces to $\frac{8}{9}$.

21. List the factors of the numerator and the denominator. The factors of 30 are 1, 2, 3, 5, **6**, 10, 15, and 30, and the factors of 48 are 1, 2, 3, 4, **6**, 8, 12, 16, 24 and 48. The greatest common factor is 6. $\frac{30}{6} = 5$ and $\frac{48}{6} = 8$. $\frac{30}{48}$ reduces to $\frac{5}{8}$.

22. List the factors of the numerator and the denominator. The factors of 56 are 1, 2, 4, 7, **8**, 14, 28, and 56, and the factors of 72 are 1, 2, 3, 4, 6, **8**, 9, 12, 18, 24, 36 and 72. The greatest common factor is 8. $\frac{56}{8} = 7$ and $\frac{72}{8} = 9$. $\frac{56}{72}$ reduces to $\frac{7}{9}$.

Fraction Basics

WHAT'S AROUND THE BEND

- ➤ Adding Like Fractions
- ➤ Adding Unlike Fractions
- ➤ Subtracting Like Fractions
- ➤ Subtracting Unlike Fractions
- ➤ Multiplying Fractions
- ➤ Dividing Fractions
- ➤ Reciprocals

In this chapter, we'll look at how to handle the four major operations—addition, subtraction, multiplication, and division—with like and unlike fractions.

ADDING LIKE FRACTIONS

Remember, like fractions are fractions that have the same denominator. To add two like fractions, add the numerators of the fractions. The denominator of your answer is the same as the denominator of the two fractions that you are adding.

Example

$$\frac{1}{5} + \frac{3}{5}$$

In this example, the two fractions are like fractions, because they both have a denominator of 5. The denominator of our answer will be 5. Add the numerators: $1 + 3 = 4$. The numerator of our answer is 4. $\frac{1}{5} + \frac{3}{5} = \frac{4}{5}$.

Example

$$\frac{6}{12} + \frac{4}{12}$$

The two fractions have a denominator of 12, so the denominator of our answer will be 12. Add the numerators: $6 + 4 = 10$. The numerator of our answer is 10. $\frac{6}{12} + \frac{4}{12} = \frac{10}{12}$. Because many tests ask for answers to be put in simplest form, let's reduce this fraction. The greatest common factor of 10 and 12 is 2. $\frac{10}{2} = 5$ and $\frac{12}{2} = 6$. $\frac{6}{12} + \frac{4}{12} = \frac{10}{12} = \frac{5}{6}$.

Example

$$\frac{1}{9} + \frac{2}{9} + \frac{3}{9}$$

These fractions have a denominator of 9, so the denominator of our answer will be 9. Add the numerators: $1 + 2 + 3 = 6$. The numerator of our answer is 6. $\frac{1}{9} + \frac{2}{9} + \frac{3}{9} = \frac{6}{9}$. The greatest common factor of 6 and 9 is 3: $\frac{6}{3} = 2$ and $\frac{9}{3} = 3$. $\frac{1}{9} + \frac{2}{9} + \frac{3}{9} = \frac{6}{9} = \frac{2}{3}$.

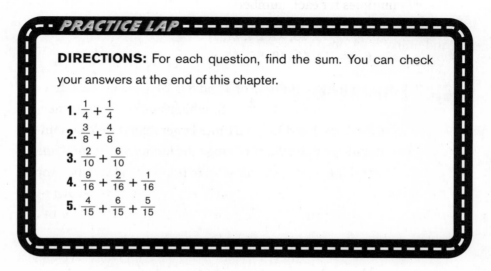

PRACTICE LAP

DIRECTIONS: For each question, find the sum. You can check your answers at the end of this chapter.

1. $\frac{1}{4} + \frac{1}{4}$

2. $\frac{3}{8} + \frac{4}{8}$

3. $\frac{2}{10} + \frac{6}{10}$

4. $\frac{9}{16} + \frac{2}{16} + \frac{1}{16}$

5. $\frac{4}{15} + \frac{6}{15} + \frac{5}{15}$

WHEN ADDING TWO of the same fraction, keep the numerator of either fraction and divide the denominator of either fraction by 2. For example, $\frac{3}{10} + \frac{3}{10} = \frac{3}{5}$. If the numerators of each addend are already in simplest form, then by adding this way, your answer will also be in simplest form. Let's check our answer: $\frac{3}{10} + \frac{3}{10} = \frac{6}{10}$. The greatest common factor of 6 and 10 is 2: $\frac{6}{2} = 3$ and $\frac{10}{2} = 5$; $\frac{3}{10} + \frac{3}{10} = \frac{6}{10} = \frac{3}{5}$. If possible, reduce fractions *before* adding them. Be sure that after you reduce them, the fractions are still like fractions!

ADDING UNLIKE FRACTIONS

This title is a little misleading. We NEVER add unlike fractions. Instead, we find common denominators and convert the unlike fractions into like fractions. You already know how to add those! Remember, to find a common denominator for two unlike fractions, you must find the least common multiple of the two denominators.

Example

$$\frac{1}{3} + \frac{2}{8}$$

First, we must find a common denominator for these fractions. List a few multiples for each number:

3: 3, 6, 9, 12, 15, 18, 21, **24**, 27, . . .
8: 8, 16, **24**, 32, 40, 48, 56, . . .

The least common multiple of 3 and 8 is 24. Convert each fraction to a number over 24: $\frac{24}{3} = 8$, which means that the new denominator of the fraction $\frac{1}{3}$ is 8 times larger than the old denominator. Because we must always change the numerator in the same way that we change the denominator, multiply the old numerator by 8: $1 \times 3\,8 = 8$. $\frac{1}{3} = \frac{8}{24}$. Now let's convert $\frac{2}{8}$: $\frac{24}{8} = 3$, which means that the new denominator of $\frac{2}{8}$ is 3 times larger than the old denominator. Multiply the old numerator by 3: $2 \times 3 = 6$. $\frac{2}{8} = \frac{6}{24}$. Now we have like fractions: $\frac{8}{24} + \frac{6}{24}$. Add the numerators and keep the denominator: $8 + 6 = 14$, so $\frac{1}{3} + \frac{2}{8} = \frac{8}{24} + \frac{6}{24} = \frac{14}{24}$. Finally, let's simplify our answer. The greatest common factor of 14 and 24 is 2: $\frac{14}{2} = 7$ and $\frac{24}{2} = 12$. $\frac{14}{24} = \frac{7}{12}$.

Example

$$\frac{3}{5} + \frac{1}{6}$$

Find a common denominator for these fractions.

5: 5, 10, 15, 20, 25, **30**, 35, . . .

6: 6, 12, 18, 24, **30**, 36, 42, . . .

The least common multiple of 5 and 6 is 30. Convert $\frac{3}{5}$ to a number over 30: $\frac{30}{5} = 6$, so we must multiply the numerator of $\frac{3}{5}$ by 6: $3 \times 6 = 18$; $\frac{3}{5} = \frac{18}{30}$. Convert $\frac{1}{6}$ in the same way: $\frac{30}{6} = 5$. Multiply the numerator of $\frac{1}{6}$ by 5: $1 \times 5 = 5$. $\frac{1}{6} = \frac{5}{30}$. Now we have like fractions: $\frac{18}{30} + \frac{5}{30}$. Add the numerators and keep the denominator: $18 + 5 = 23$, so $\frac{3}{5} + \frac{1}{6} = \frac{18}{30} + \frac{5}{30} = \frac{23}{30}$.

Example

$$\frac{2}{9} + \frac{5}{12} + \frac{1}{3}$$

Find a common denominator for these fractions.

9: 9, 18, 27, **36**, 45, 54, 63, . . .

12: 12, 24, **36**, 48, 60, 72, 84, . . .

3: 3, 6, 9, 12, 15, 18, 21, 24, 27, 30, 33, **36**, 39, . . .

The least common multiple of 9, 12, and 3 is 36. Convert each fraction to a number over 36:

$$\frac{2}{9}: \frac{36}{9} = 4, 4 \times 2 = 8. \frac{2}{9} = \frac{8}{36}.$$

$$\frac{5}{12}: \frac{36}{12} = 3, 3 \times 5 = 15. \frac{5}{12} = \frac{15}{36}.$$

$$\frac{1}{3}: \frac{36}{3} = 12, 12 \times 1 = 12. \frac{1}{3} = \frac{12}{36}.$$

Now we have like fractions: $\frac{8}{36} + \frac{15}{36} + \frac{12}{36}$. Add the numerators and keep the denominator: $8 + 15 + 12 = 35$, so $\frac{2}{9} + \frac{5}{12} + \frac{1}{3} = \frac{8}{36} + \frac{15}{36} + \frac{12}{36} = \frac{35}{36}$.

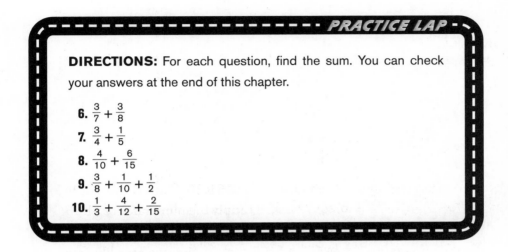

PRACTICE LAP

DIRECTIONS: For each question, find the sum. You can check your answers at the end of this chapter.

6. $\frac{3}{7} + \frac{3}{8}$

7. $\frac{3}{4} + \frac{1}{5}$

8. $\frac{4}{10} + \frac{6}{15}$

9. $\frac{3}{8} + \frac{1}{10} + \frac{1}{2}$

10. $\frac{1}{3} + \frac{4}{12} + \frac{2}{15}$

SUBTRACTING LIKE FRACTIONS

To add like fractions, we add the numerators and keep the denominators. You can probably guess how to subtract like fractions: Subtract the numerator of the second fraction from the numerator of the first, and keep the denominators.

Example

$\frac{7}{11} - \frac{4}{11}$

Both fractions have a denominator of 11, so the denominator of our answer will be 11. Subtract the numerator of the second fraction from the numerator of the first fraction: $7 - 4 = 3$. The numerator of our answer is 3: $\frac{7}{11} - \frac{4}{11} = \frac{3}{11}$.

Example

$$\frac{5}{8} - \frac{3}{8}$$

The denominator of our answer will be 8. Because $5 - 3 = 2$, $\frac{5}{8} - \frac{3}{8} = \frac{2}{8}$. The greatest common factor of 2 and 8 is 2, so $\frac{2}{8} = \frac{1}{4}$.

Example

$$\frac{11}{12} - \frac{5}{12} - \frac{1}{12}$$

These fractions have a denominator of 12, so the denominator of our answer will be 12. $11 - 5 - 1 = 5$. The numerator of our answer is 5: $\frac{11}{12} - \frac{5}{12} - \frac{1}{12} = \frac{5}{12}$.

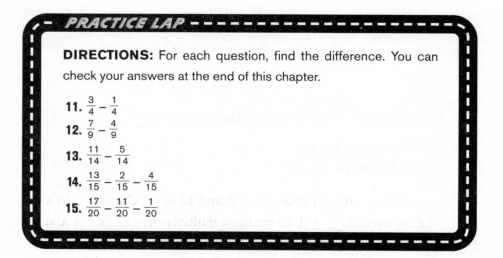

PRACTICE LAP

DIRECTIONS: For each question, find the difference. You can check your answers at the end of this chapter.

11. $\frac{3}{4} - \frac{1}{4}$

12. $\frac{7}{9} - \frac{4}{9}$

13. $\frac{11}{14} - \frac{5}{14}$

14. $\frac{13}{15} - \frac{2}{15} - \frac{4}{15}$

15. $\frac{17}{20} - \frac{11}{20} - \frac{1}{20}$

SUBTRACTING UNLIKE FRACTIONS

Just like with addition, we never subtract unlike fractions. We find common denominators, and then subtract.

Example

$$\frac{2}{3} - \frac{3}{5}$$

We begin by finding a common denominator for 3 and 5. List a few multiples for each number:

  3: 3, 6, 9, 12, **15**, 18, . . .
  5: 5, 10, **15**, 20, 25, . . .

The least common multiple of 3 and 5 is 15. Convert $\frac{2}{3}$ to a number over 15. $\frac{15}{3} = 5$. Multiply the numerator and denominator of $\frac{2}{3}$ by 5: $2 \times 5 = 10$. $\frac{2}{3} = \frac{10}{15}$. Now convert $\frac{3}{5}$ to a number over 15. $\frac{15}{5} = 3$. Multiply the numerator and denominator of $\frac{3}{5}$ by 3: $3 \times 3 = 9$; $\frac{3}{5} = \frac{9}{15}$. Now we have like fractions: $\frac{10}{15} - \frac{9}{15}$. Find the difference between 10 and 9 and keep the denominator: $10 - 9 = 1$, so $\frac{2}{3} - \frac{3}{5} = \frac{10}{15} - \frac{9}{15} = \frac{1}{15}$.

Example

$$\frac{7}{8} - \frac{7}{12}$$

Find a common denominator for these fractions.

  8: 8, 16, **24**, 32, 40, . . .
  12: 12, **24**, 36, 48, 60, . . .

The least common multiple of 8 and 12 is 24. Convert $\frac{7}{8}$ to a number over 24. $\frac{24}{8} = 3$, so we must multiply the numerator and denominator of $\frac{7}{8}$ by 3: $7 \times 3 = 21$; $\frac{7}{8} = \frac{21}{24}$. Convert $\frac{7}{12}$ to a number over 24. $\frac{24}{12} = 2$, so we must multiply the numerator and denominator of $\frac{7}{12}$ by 2: $7 \times 2 = 14$. $\frac{7}{12} = \frac{14}{24}$. Now we have like fractions: $\frac{21}{24} - \frac{14}{24}$. Find the difference between 21 and 14 and keep the denominator: $21 - 14 = 7$, so $\frac{7}{8} - \frac{7}{12} = \frac{21}{24} - \frac{14}{24} = \frac{7}{24}$.

Example

$$\frac{5}{6} - \frac{1}{2} - \frac{1}{12}$$

Find a common denominator for these fractions.

6: 6, **12**, 18, 24, . . .

2: 2, 4, 6, 8, 10, **12**, 14, . . .

12: **12**, 24, 36, 48, . . .

The least common multiple of 6, 2, and 12 is 12. Convert each fraction to a number over 12:

$$\frac{5}{6} \colon \frac{12}{6} = 2, 5 \times 2 = 10; \frac{5}{6} = \frac{10}{12}.$$
$$\frac{1}{2} \colon \frac{12}{2} = 6, 1 \times 6 = 6; \frac{1}{2} = \frac{6}{12}.$$

Now we have like fractions: $\frac{10}{12} - \frac{6}{12} - \frac{1}{12}$. Because $10 - 6 - 1 = 3$, $\frac{10}{12} - \frac{6}{12} - \frac{1}{12} = \frac{3}{12}$. The greatest common factor of 3 and 12 is 3, so the numerator of our answer reduces to $\frac{3}{3} = 1$ and the denominator reduces to $\frac{12}{3} = 4$. $\frac{3}{12} = \frac{1}{4}$.

PRACTICE LAP

DIRECTIONS: For each question, find the difference. You can check your answers at the end of this chapter.

16. $\frac{13}{16} - \frac{5}{8}$

17. $\frac{5}{7} - \frac{4}{9}$

18. $\frac{17}{20} - \frac{2}{3}$

19. $\frac{15}{16} - \frac{1}{3} - \frac{1}{6}$

20. $\frac{9}{10} - \frac{1}{8} - \frac{2}{5}$

PACE YOURSELF

ASK TEN PEOPLE what kind of pets they have (dogs, cats, fish, etc.). What fraction of them has dogs or cats? In other words, add the fraction of people that have dogs to the fraction of people that have cats. Find the fraction that represents the amount of people who have at least one pet. What other questions about pets can you answer by adding and subtracting fractions?

MULTIPLYING FRACTIONS

Good news—you don't need common denominators to multiply fractions! You can multiply two fractions whether they are like or unlike. We do not need common denominators because the denominator of our answer does not have to be the same as the denominators of the two fractions that we are multiplying.

To multiply two fractions, multiply the numerators, and then multiply the denominators. Multiplying fractions is actually easier than adding or subtracting them!

Example

$$\frac{3}{7} \times \frac{4}{9}$$

First, multiply the numerators: $3 \times 4 = 12$. Then, multiply the denominators: $7 \times 9 = 63$. The product of $\frac{3}{7}$ and $\frac{4}{9}$ is $\frac{12}{63}$. The greatest common factor of 12 and 63 is 3, so $\frac{12}{63}$ reduces to $\frac{4}{21}$.

Example

$$\frac{7}{10} \times \frac{5}{6}$$

Multiply the numerators: $7 \times 5 = 35$. Multiply the denominators: $10 \times 6 = 60$. The product of $\frac{7}{10}$ and $\frac{5}{6}$ is $\frac{35}{60}$. The greatest common factor of 35 and 60 is 5, so $\frac{35}{60}$ reduces to $\frac{7}{12}$.

INSIDE TRACK

YOU'VE SEEN HOW to reduce a single fraction by dividing the numerator and denominator of that fraction by the same number. When multiplying fractions, we can divide the numerator of one fraction and the denominator of the other fraction by the same number. Look again at the last example. Before multiplying, we can divide the numerator of $\frac{5}{6}$ by 5 and the denominator of $\frac{7}{10}$ by 5, because 5 is the greatest common factor of 5 and 10. The problem now becomes $\frac{7}{2} \times \frac{1}{6}$, which is easier to multiply. The product of these fractions is $\frac{7}{12}$, which is the same as the product of $\frac{7}{10} \times \frac{5}{6}$—and the answer is already in simplest form!

Example

$$\frac{3}{11} \times \frac{2}{3}$$

Before multiplying, look at the numerator of the first fraction and the denominator of the second fraction—they are the same. You can divide both by 3, or "cancel" the threes. The problem now becomes $\frac{1}{11} \times \frac{2}{1}$. Multiply the numerators: $1 \times 2 = 2$. Multiply the denominators: $11 \times 1 = 11$. $\frac{3}{11} \times \frac{2}{3} = \frac{1}{11} \times \frac{2}{1} = \frac{2}{11}$.

Sometimes you can divide the numerators and denominators more than once before multiplying.

Example

$$\frac{6}{15} \times \frac{5}{14}.$$

We could begin with any part of this problem—we could simplify the first fraction, divide the numerator of the first fraction and the denominator of the second fraction, or divide the denominator of the first fraction and the numerator of the second fraction.

Let's start by dividing the denominator of the first fraction and the numerator of the second fraction by 5, because 5 is the greatest common factor of 5 and 15. Because $\frac{15}{5} = 3$ and $\frac{5}{5} = 1$, the problem becomes $\frac{6}{3} \times \frac{1}{14}$. Next, let's divide the numerator and the denominator of the first fraction by 3, because 3 is the greatest common factor of 3 and 6. Because $\frac{6}{3} = 2$ and $\frac{3}{3} = 1$, the problem becomes: $\frac{2}{1} \times \frac{1}{14}$. Finally, let's divide the numerator of the first fraction and the denominator of the second fraction by 2, because 2 is the greatest common factor of 2 and 14. $\frac{2}{2} = 1$ and $\frac{14}{2} = 7$. We have reduced the problem to $\frac{1}{1} \times \frac{1}{7}$, or $1 \times \frac{1}{7}$, which is $\frac{1}{7}$. After all that simplifying, multiplication was a snap!

CAUTION!

YOU CAN DIVIDE the numerator and denominator of the same fraction by the same number, and you can divide the numerator of one fraction and the denominator of another fraction by the same number before multiplying. However, you CANNOT divide the numerator of one fraction and the numerator of another fraction by the same number before multiplying. You also cannot divide the denominator of one fraction and the denominator of another fraction by the same number before multiplying. The problem $\frac{1}{3} \times \frac{1}{3}$ cannot be simplified before multiplying.

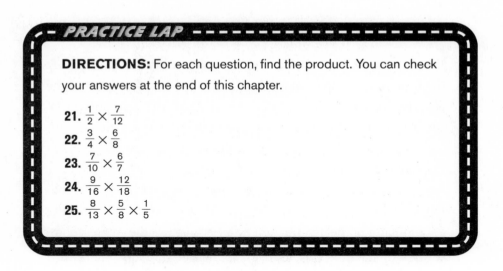

PRACTICE LAP

DIRECTIONS: For each question, find the product. You can check your answers at the end of this chapter.

21. $\frac{1}{2} \times \frac{7}{12}$

22. $\frac{3}{4} \times \frac{6}{8}$

23. $\frac{7}{10} \times \frac{6}{7}$

24. $\frac{9}{16} \times \frac{12}{18}$

25. $\frac{8}{13} \times \frac{5}{8} \times \frac{1}{5}$

FRACTIONS AS DIVISION

It's easy to forget that the fraction bar in a fraction actually represents division. The fraction $\frac{3}{4}$ means "3 divided by 4." That is why we can simplify fractions such as $\frac{8}{2}$ into 4: because $\frac{8}{2}$ means "8 divided by 2."

DIVIDING FRACTIONS

As with multiplying fractions, we do not need common denominators to divide one fraction by another fraction. Here's what's really strange: When we divide fractions, we don't even use division—we use multiplication! What's the catch? Before multiplying, we find the **reciprocal** of the divisor. The reciprocal of a fraction is easy to find—just flip it over.

The **reciprocal** of a fraction is the multiplicative inverse of the fraction. That's a complicated way of saying switch the numerator and the denominator. The reciprocal of $\frac{3}{4}$ is $\frac{4}{3}$. The numerator and the denominator switch places. $\frac{4}{3}$ is the multiplicative inverse because $\frac{3}{4} \times \frac{4}{3} = 1$. Any number times its reciprocal is equal to 1—even whole numbers. Whole numbers have a denominator of 1. The number 8 can also be written as $\frac{8}{1}$. The reciprocal of 8 is $\frac{1}{8}$: $8 \times \frac{1}{8} = 1$.

Before we divide fractions, let's look at division of whole numbers.

Example

$8 \div 2$

In this example, 8 is the dividend and 2 is the divisor. The dividend always comes before the division symbol, and the divisor always comes after the division symbol: $8 \div 2 = 4$.

Now that we've reviewed the names of the parts of a division problem, let's divide some fractions!

Example

$\frac{1}{4} \div \frac{5}{3}$

In this example, $\frac{1}{4}$ is the dividend and $\frac{5}{3}$ is the divisor. The first step in solving a fraction division problem is to find the reciprocal of the divisor. Switch the numerator and denominator of the fraction $\frac{5}{3}$; its reciprocal is $\frac{3}{5}$. Next, switch the division symbol to a multiplication symbol: $\frac{1}{4} \div \frac{5}{3}$ becomes $\frac{1}{4} \times \frac{3}{5}$—these two math problems have exactly the same answer! You know what to do next: Multiply the numerators and multiply the denominators: $1 \times 3 = 3$ and $4 \times 5 = 20$, which means that $\frac{1}{4} \times \frac{3}{5}$, and $\frac{1}{4} \div \frac{5}{3}$, both equal $\frac{3}{20}$.

Example

$$\frac{8}{9} \div \frac{4}{9}$$

This problem can be solved two ways—in fact, all fraction division problems can be solved two ways. The easiest method is the one we just saw—take the reciprocal of the divisor, and multiply. However, if you already have common denominators, you can simply divide the numerator of the dividend by the numerator of the divisor. Because $8 \div 4 = 2$, $\frac{8}{9} \div \frac{4}{9} = 2$. Now let's use the first method to check our answer. Because $\frac{4}{9}$ is the divisor, find its reciprocal. Switching the numerator and the denominator, we find that its reciprocal is $\frac{9}{4}$. Next, switch the division symbol to a multiplication symbol. $\frac{8}{9} \div \frac{4}{9}$ becomes $\frac{8}{9} \times \frac{9}{4}$. We can cancel the 9 in the denominator of $\frac{8}{9}$ with the 9 in the numerator of $\frac{9}{4}$. Now the problem becomes $\frac{8}{1} \times \frac{1}{4}$. Multiply the numerators and multiply the denominators. $8 \times 1 = 8$ and $1 \times 4 = 4$, which means that $\frac{8}{1} \times \frac{1}{4}$, and $\frac{8}{9} \div \frac{4}{9}$, equal $\frac{8}{4}$, or 2. Both methods give us the same answer.

CAUTION!

BEFORE YOU CAN cancel or divide the numerators and denominators of the fractions in a division problem, you must first take the reciprocal of the divisor and switch the division symbol to a multiplication symbol. For instance, in the problem $\frac{2}{3} \div \frac{3}{7}$, you cannot cancel the 3 in the denominator of $\frac{2}{3}$ with the 3 in the numerator of $\frac{3}{7}$. The reciprocal of $\frac{3}{7}$ is $\frac{7}{3}$; $\frac{2}{3} \div \frac{3}{7}$ becomes $\frac{2}{3} \times \frac{7}{3}$, and you can see now that neither fraction can be simplified.

Example

$$\frac{5}{21} \div \frac{10}{18}$$

Because these are not like fractions, we'll solve this problem using the first method. The reciprocal of $\frac{10}{18}$ is $\frac{18}{10}$, so $\frac{5}{21} \div \frac{10}{18}$ becomes $\frac{5}{21}$ $\times \frac{18}{10}$. Now that it is a multiplication problem, we can simplify these fractions. By dividing the numerator of $\frac{5}{21}$ and the denominator of $\frac{18}{10}$ by 5, and by dividing the denominator of $\frac{5}{21}$ and the numerator of $\frac{18}{10}$ by 3, $\frac{5}{21} \times \frac{18}{10}$ becomes $\frac{1}{7} \times \frac{6}{2}$. $\frac{6}{2} = 3$, so now the problem becomes $\frac{1}{7} \times 3$, which is equal to $\frac{3}{7}$.

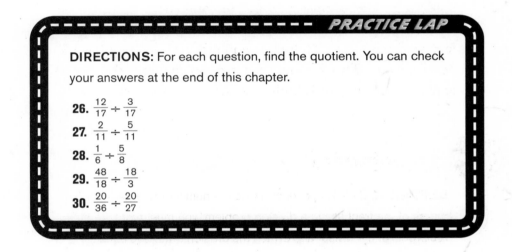

PRACTICE LAP

DIRECTIONS: For each question, find the quotient. You can check your answers at the end of this chapter.

26. $\frac{12}{17} \div \frac{3}{17}$

27. $\frac{2}{11} \div \frac{5}{11}$

28. $\frac{1}{6} \div \frac{5}{8}$

29. $\frac{48}{18} \div \frac{18}{3}$

30. $\frac{20}{36} \div \frac{20}{27}$

Now we've mastered the four major operations (addition, subtraction, multiplication, and division) with like and unlike fractions. Up next: the same operations—with mixed numbers.

ANSWERS

1. The denominators of these fractions are 4, so the denominator of our answer will be 4. The sum of the numerators is $1 + 1 = 2$: $\frac{1}{4} + \frac{1}{4} = \frac{2}{4}$. The greatest common factor of 2 and 2 is 2. $\frac{2}{2} = 1$ and $\frac{4}{2} = 2$. $\frac{1}{4} + \frac{1}{4} = \frac{2}{4} = \frac{1}{2}$.

2. The denominators of these fractions are 8, so the denominator of our answer will be 8. The sum of the numerators is $3 + 4 = 7$: $\frac{3}{8} + \frac{4}{8} = \frac{7}{8}$.

3. The denominators of these fractions are 10, so the denominator of our answer will be 10. The sum of the numerators is $2 + 6 = 8$: $\frac{2}{10} + \frac{6}{10} = \frac{8}{10}$. The greatest common factor of 8 and 10 is 2. $\frac{8}{2} = 4$ and $\frac{10}{2} = 5$. $\frac{2}{10} + \frac{6}{10} = \frac{8}{10} = \frac{4}{5}$.

4. The denominators of these fractions are 16, so the denominator of our answer will be 16. The sum of the numerators is $9 + 2 + 1 = 12$: $\frac{9}{16} + \frac{2}{16} + \frac{1}{16} = \frac{12}{16}$. The greatest common factor of 12 and 16 is 4: $\frac{12}{4} = 3$ and $\frac{16}{4} = 4$. $\frac{9}{16} + \frac{2}{16} + \frac{1}{16} = \frac{12}{16} = \frac{3}{4}$.

5. The denominators of these fractions are 15, so the denominator of our answer will be 15. The sum of the numerators is $4 + 6 + 5 = 15$: $\frac{4}{15} + \frac{6}{15} + \frac{5}{15} = \frac{15}{15}$. Any number over itself is equal to 1, so $\frac{4}{15} + \frac{6}{15} + \frac{5}{15} = \frac{15}{15} = 1$.

6. The least common multiple of 7 and 8 is 56. Multiply the numerator and denominator of $\frac{3}{7}$ by 8: $\frac{3}{7} = \frac{24}{56}$. Multiply the numerator and denominator of $\frac{3}{8}$ by 7: $\frac{3}{8} = \frac{21}{56}$. The sum of the numerators is $24 + 21 = 45$: $\frac{3}{7} + \frac{3}{8} = \frac{24}{56} + \frac{21}{56} = \frac{45}{56}$.

7. The least common multiple of 4 and 5 is 20. Multiply the numerator and denominator of $\frac{3}{4}$ by 5: $\frac{3}{4} = \frac{15}{20}$. Multiply the numerator and denominator of $\frac{1}{5}$ by 4: $\frac{1}{5} = \frac{4}{20}$. The sum of the numerators is $15 + 4 = 19$: $\frac{3}{4} + \frac{1}{5} = \frac{15}{20} + \frac{4}{20} = \frac{19}{20}$.

8. The least common multiple of 10 and 15 is 30. Multiply the numerator and denominator of $\frac{4}{10}$ by 3: $\frac{4}{10} = \frac{12}{30}$. Multiply the numerator and denominator of $\frac{6}{15}$ by 2: $\frac{6}{15} = \frac{12}{30}$. The sum of the numerators is $12 + 12 = 24$: $\frac{4}{10} + \frac{6}{15} = \frac{12}{30} + \frac{12}{30} = \frac{24}{30}$. The greatest common factor of 24 and 30 is 6. $\frac{24}{6} = 4$ and $\frac{30}{6} = 5$; $\frac{24}{30} = \frac{4}{5}$.

9. The least common multiple of 8, 10, and 2 is 40. Multiply the numerator and denominator of $\frac{3}{8}$ by 5: $\frac{3}{8} = \frac{15}{40}$. Multiply the numerator and denominator of $\frac{1}{10}$ by 4: $\frac{1}{10} = \frac{4}{40}$. Multiply the numerator and denominator of $\frac{1}{2}$ by 20: $\frac{1}{2} = \frac{20}{40}$. The sum of the numerators is $15 + 4 + 20 = 39$: $\frac{3}{8} + \frac{1}{10} + \frac{1}{2} = \frac{15}{40} + \frac{4}{40} + \frac{20}{40} = \frac{39}{40}$.

10. The least common multiple of 3, 12, and 15 is 60. Multiply the numerator and denominator of $\frac{1}{3}$ by 20: $\frac{1}{3} = \frac{20}{60}$. Multiply the numerator and denominator of $\frac{4}{12}$ by 5: $\frac{4}{12} = \frac{20}{60}$. Multiply the numerator and denominator of $\frac{2}{15}$ by 4: $\frac{2}{15} = \frac{8}{60}$. The sum of the numerators is $20 + 20 + 8 = 48$: $\frac{1}{3} + \frac{4}{12} + \frac{2}{15} = \frac{20}{60} + \frac{20}{60} + \frac{8}{60} = \frac{48}{60}$. The greatest common factor of 48 and 60 is 12. $\frac{48}{12} = 4$ and $\frac{60}{12} = 5$. $\frac{48}{60} = \frac{4}{5}$.

11. The denominators of these fractions are 4, so the denominator of our answer will be 4. Because $3 - 1 = 2$, $\frac{3}{4} - \frac{1}{4} = \frac{2}{4}$. The greatest common factor of 2 and 4 is 2. $\frac{2}{2} = 1$ and $\frac{4}{2} = 2$. $\frac{3}{4} - \frac{1}{4} = \frac{2}{4} = \frac{1}{2}$.

12. The denominators of these fractions are 9, so the denominator of our answer will be 9. Because $7 - 4 = 3$, $\frac{7}{9} - \frac{4}{9} = \frac{3}{9}$. The greatest common factor of 3 and 9 is 3. $\frac{3}{3} = 1$ and $\frac{9}{3} = 3$: $\frac{7}{9} - \frac{4}{9} = \frac{3}{9} = \frac{1}{3}$.

13. The denominators of these fractions are 14, so the denominator of our answer will be 14. Because $11 - 5 = 6$, $\frac{11}{14} - \frac{5}{14} = \frac{6}{14}$. The greatest common factor of 6 and 14 is 2. $\frac{6}{2} = 3$ and $\frac{14}{2} = 7$: $\frac{11}{14} - \frac{5}{14} = \frac{6}{14} = \frac{3}{7}$.

14. The denominators of these fractions are 15, so the denominator of our answer will be 15. Because $13 - 2 - 4 = 7$, $\frac{13}{15} - \frac{2}{15} - \frac{4}{15} = \frac{7}{15}$.

15. The denominators of these fractions are 20, so the denominator of our answer will be 20. Because $17 - 11 - 1 = 5$, $\frac{17}{20} - \frac{11}{20} - \frac{1}{20} = \frac{5}{20}$. The greatest common factor of 5 and 20 is 5: $\frac{5}{5} = 1$ and $\frac{20}{5} = 4$; $\frac{17}{20} - \frac{11}{20} - \frac{1}{20} = \frac{5}{20} = \frac{1}{4}$.

16. The least common multiple of 16 and 8 is 16. Multiply the numerator and denominator of $\frac{5}{8}$ by 2: $\frac{5}{8} = \frac{10}{16}$. Subtract the second numerator from the first: $13 - 10 = 3$: $\frac{13}{16} - \frac{5}{8} = \frac{13}{16} - \frac{10}{16} = \frac{3}{16}$.

17. The least common multiple of 7 and 9 is 63. Multiply the numerator and denominator of $\frac{5}{7}$ by 9: $\frac{5}{7} = \frac{45}{63}$. Multiply the numerator and denominator of $\frac{4}{9}$ by 7: $\frac{4}{9} = \frac{28}{63}$. Subtract the second numerator from the first: $45 - 28 = 17$. $\frac{5}{7} - \frac{4}{9} = \frac{45}{63} - \frac{28}{63} = \frac{17}{63}$.

18. The least common multiple of 20 and 3 is 60. Multiply the numerator and denominator of $\frac{17}{20}$ by 3: $\frac{17}{20} = \frac{51}{60}$. Multiply the numerator and denominator of $\frac{2}{3}$ by 20: $\frac{2}{3} = \frac{40}{60}$. Subtract the second numerator from the first: $51 - 40 = 11$: $\frac{17}{20} - \frac{2}{3} = \frac{51}{60} - \frac{40}{60} = \frac{11}{60}$.

19. The least common multiple of 16, 3 and 6 is 48. Multiply the numerator and denominator of $\frac{15}{16}$ by 3: $\frac{15}{16} = \frac{45}{48}$. Multiply the numerator and denominator of $\frac{1}{3}$ by 16: $\frac{1}{3} = \frac{16}{48}$. Multiply the numerator and denominator of $\frac{1}{6}$ by 8: $\frac{1}{6} = \frac{8}{48}$. Subtract the second numerator and the third numerator from the first: $45 - 16 - 8 = 21$: $\frac{15}{16} - \frac{1}{3} - \frac{1}{6} = \frac{45}{48} - \frac{16}{48} - \frac{8}{48}$ $= \frac{21}{48}$. The greatest common factor of 21 and 48 is 3: $\frac{21}{3} = 7$ and $\frac{48}{3} = 16$; $\frac{21}{48} = \frac{7}{16}$.

20. The least common multiple of 10, 8 and 5 is 40. Multiply the numerator and denominator of $\frac{9}{10}$ by 4: $\frac{9}{10} = \frac{36}{40}$. Multiply the numerator and denominator of $\frac{1}{8}$ by 5: $\frac{1}{8} = \frac{5}{40}$. Multiply the numerator and denominator of $\frac{2}{5}$ by 8: $\frac{2}{5} = \frac{16}{40}$. Subtract the second numerator and the third numerator from the first: $36 - 5 - 16 = 15$: $\frac{9}{10} - \frac{1}{8} - \frac{2}{5} = \frac{36}{40} - \frac{5}{40} - \frac{16}{40}$ $= \frac{15}{40}$. The greatest common factor of 15 and 40 is 5. $\frac{15}{5} = 3$ and $\frac{40}{5} =$ 8. $\frac{15}{40} = \frac{3}{8}$.

21. $1 \times 7 = 7$ and $2 \times 12 = 24$, so $\frac{1}{2} \times \frac{7}{12} = \frac{7}{24}$.

22. $3 \times 6 = 18$ and $4 \times 8 = 32$, so $\frac{3}{4} \times \frac{6}{8} = \frac{18}{32}$, which reduces to $\frac{9}{16}$. We also could have divided the numerator and denominator of $\frac{6}{8}$ by 2 before multiplying.

23. $7 \times 6 = 42$ and $10 \times 7 = 70$, so $\frac{7}{10} \times 7 = \frac{42}{70}$, which reduces to $\frac{3}{5}$. We also could have canceled the 7 in the numerator of $\frac{7}{10}$ with the 7 in the denominator of $\frac{6}{7}$, and divided the denominator of $\frac{7}{10}$ and the numerator of $\frac{6}{7}$ by 2 before multiplying.

24. $9 \times 12 = 108$ and $16 \times 18 = 288$, so $\frac{9}{16} \times \frac{12}{8} = \frac{108}{288}$, which reduces to $\frac{3}{8}$. We also could have divided the denominator of $\frac{9}{16}$ and the numerator of $\frac{12}{18}$ by 4 and divided the numerator of $\frac{9}{16}$ and the denominator of $\frac{12}{18}$ by 9 before multiplying.

25. $8 \times 5 \times 1 = 40$ and $13 \times 8 \times 5 = 520$, so $\frac{8}{13} \times \frac{5}{8} \times \frac{1}{5} = \frac{40}{520}$, which reduces to $\frac{1}{13}$. We also could have canceled the 8 in the numerator of $\frac{8}{13}$ with the 8 in the denominator of $\frac{5}{8}$ and canceled the 5 in the numerator of $\frac{5}{8}$ with the 5 in the denominator of $\frac{1}{5}$ before multiplying.

26. Because we have common denominators, divide the numerator of the first fraction by the numerator of the second fraction: $\frac{12}{3} = 4$.

27. Because we have common denominators, divide the numerator of the first fraction by the numerator of the second fraction: $\frac{2}{5} = \frac{2}{5}$.

28. The reciprocal of the divisor, $\frac{5}{8}$, is $\frac{8}{5}$; $\frac{1}{6} \div \frac{5}{8}$ becomes $\frac{1}{6} \times \frac{8}{5}$. Divide the 6 in $\frac{1}{6}$ and the 8 in $\frac{8}{5}$ by 2, and the problem becomes $\frac{1}{3} \times \frac{4}{5}$. Multiply the numerators and the denominators: $1 \times 4 = 4$ and $3 \times 5 = 15$, so $\frac{1}{3} \times \frac{4}{5}$, and $\frac{1}{6} \div \frac{5}{8}$, equal $\frac{4}{15}$.

29. The reciprocal of the divisor, $\frac{18}{3}$, is $\frac{3}{18}$; $\frac{48}{18} \div \frac{18}{3}$ becomes $\frac{48}{18} \times \frac{3}{18}$. Divide the 18 in $\frac{48}{18}$ and the 3 in $\frac{3}{18}$ by 3, and the problem becomes $\frac{48}{6} \times \frac{1}{18}$, or $8 \times \frac{1}{18}$, which is equal to $\frac{8}{18}$. Simplify $\frac{8}{18}$ by dividing the numerator and the denominator by 2: $\frac{8}{18} = \frac{4}{9}$.

30. The reciprocal of the divisor, $\frac{20}{27}$, is $\frac{27}{20}$; $\frac{20}{36} \div \frac{20}{27}$ becomes $\frac{20}{36} \times \frac{27}{20}$. Cancel the 20 in $\frac{20}{36}$ with the 20 in $\frac{27}{20}$ and divide the 36 in $\frac{20}{36}$ and the 27 in $\frac{27}{20}$ by 9, and the problem becomes $\frac{1}{4} \times \frac{3}{1}$. Multiply the numerators and the denominators: $1 \times 3 = 3$ and $4 \times 1 = 4$, so $\frac{1}{4} \times \frac{3}{1}$, and $\frac{20}{36} \div \frac{20}{27}$, equal $\frac{3}{4}$.

5

Mixed Numbers

WHAT'S AROUND THE BEND

- ➡ What's a Mixed Number?
- ➡ Converting Improper Fractions to Mixed Numbers
- ➡ Converting Mixed Numbers to Improper Fractions
- ➡ Adding Mixed Numbers
- ➡ Subtracting Mixed Numbers
- ➡ Multiplying Mixed Numbers
- ➡ Reciprocals of Mixed Numbers
- ➡ Dividing Mixed Numbers

PART WHOLE NUMBER, PART FRACTION

So far, we've looked at proper fractions—fractions that are between –1 and 1—and improper fractions—fractions that are greater than or equal to 1, or less than or equal to –1. We can rewrite improper fractions as mixed numbers.

> ### FUEL FOR THOUGHT
>
> **A MIXED NUMBER** has two parts: a whole number part and a fraction part. The number $1\frac{1}{2}$ is a mixed number. Mixed numbers, like improper fractions, have a value that is greater than or equal to 1, or less than or equal to −1. The improper fraction $\frac{3}{2}$ is equal to the mixed number $1\frac{1}{2}$.

Improper Fractions Become Mixed Numbers . . .

Remember, the fraction bar represents division. $\frac{5}{8}$ means "5 divided by 8." To convert an improper fraction to a mixed number, divide the numerator by the denominator.

Example

$\frac{7}{3}$

In this improper fraction, the numerator is 7 and the denominator is 3; 7 divided by 3 is 2, with 1 left over. The whole number part of our answer is 2, and we express the remainder as a fraction. Because the improper fraction has a denominator of 3, our remainder has a denominator of 3; 7 divided by 3 is $2\frac{1}{3}$.

Example

$\frac{11}{4}$

Divide the numerator by the denominator: 11 divided by 4 is 2, with 3 left over, which means that our mixed number is $2\frac{3}{4}$.

Sometimes, an improper fraction can be converted to a whole number. After dividing the numerator by the denominator, there is no remainder, so there is no fractional part of our answer, just a whole number.

Example

$$\frac{15}{5}$$

15 divided by 5 is 3, with no remainder. Our answer is a whole number, not a mixed number.

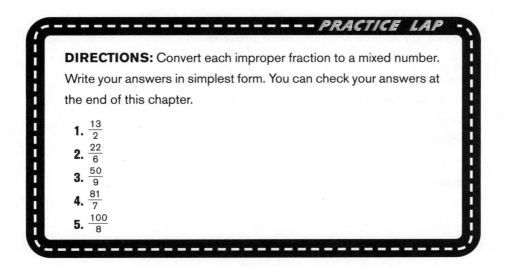

PRACTICE LAP

DIRECTIONS: Convert each improper fraction to a mixed number. Write your answers in simplest form. You can check your answers at the end of this chapter.

1. $\frac{13}{2}$

2. $\frac{22}{6}$

3. $\frac{50}{9}$

4. $\frac{81}{7}$

5. $\frac{100}{8}$

. . . And Mixed Numbers Become Improper Fractions

Just as we can convert improper fractions to mixed numbers, we can convert mixed numbers to improper fractions. This will soon become important, as we learn to add, subtract, multiply, and divide mixed numbers.

There are three steps to converting a mixed number to an improper fraction. First, multiply the whole number by the denominator of the fraction. Second, add to that product the numerator of the fraction. Third, put that sum over the denominator of the fraction.

Example

$$5\frac{3}{4}$$

We begin by multiplying the whole number, 5, by the denominator of the fraction, 4: $5 \times 4 = 20$. Next, add to that product the numerator of the fraction: $20 + 3 = 23$. Finally, put that sum over the denominator of the fraction: $5\frac{3}{4} = \frac{23}{4}$.

A WHOLE NUMBER can also be converted to an improper fraction —with any denominator. A whole number, such as 4, can be written as $\frac{4}{1}$, because any whole number can be written as itself over 1. But, in the same way that we change fractions when we find common denominators, we can multiply the numerator and denominator of $\frac{4}{1}$ by any number we want (as long as we multiply them both by the same number). If we multiply the top and bottom by 6, we get $\frac{24}{6}$. 4 $= \frac{4}{1} = \frac{24}{6}$, and there are infinitely more improper fractions we could make that are equal to 4.

Example

$8\frac{2}{9}$

Follow the three steps. First, $8 \times 9 = 72$. Next, $72 + 2 = 74$. Finally, $8\frac{2}{9} = \frac{74}{9}$.

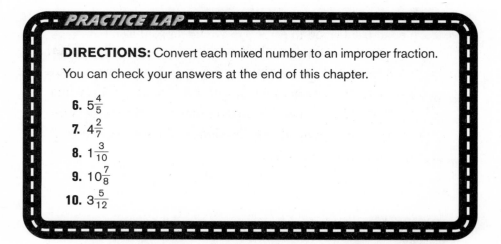

PRACTICE LAP

DIRECTIONS: Convert each mixed number to an improper fraction. You can check your answers at the end of this chapter.

6. $5\frac{4}{5}$

7. $4\frac{2}{7}$

8. $1\frac{3}{10}$

9. $10\frac{7}{8}$

10. $3\frac{5}{12}$

ADDING MIXED NUMBERS

Now let's look at adding mixed numbers. Knowing how to convert improper fractions to mixed numbers will help us simplify our answers.

To add two mixed numbers, first, add the whole number parts of each number. Then, add the fractions. If the fractions are unlike, we may need to find common denominators. Finally, if the sum of the fractions is an improper fraction, convert that fraction to a mixed number and add it to the whole number part of your answer.

Example

$$4\frac{1}{5} + 6\frac{3}{5}$$

First, add the whole number parts of each number: $4 + 6 = 10$. Next, add the fractions. Because these fractions are like, just add the numerators and keep the denominator. $\frac{1}{5} + \frac{3}{5} = \frac{4}{5}$. This sum is a proper fraction, so our answer is $10\frac{4}{5}$.

Here's an example where the sum of the fractions is an improper fraction—we'll have a little more work to do!

Example

$$3\frac{3}{4} + 1\frac{3}{4}$$

Again, begin by adding the whole number parts of each number: $3 + 1 = 4$. Next, add the fractions: $\frac{3}{4} + \frac{3}{4} = \frac{6}{4}$. Now, convert $\frac{6}{4}$ to a mixed number; 6 divided by 4 is 1, with 2 left over: $\frac{6}{4} = 1\frac{2}{4}$, which reduces to $1\frac{1}{2}$. Finally, add this mixed number to the whole number part of our answer by adding the whole number parts, and keeping the fraction: $4 + 1\frac{1}{2} = 5\frac{1}{2}$.

CAUTION!

THE SUM OF two mixed numbers may not have a fractional part. Sometimes, the fractional parts of two mixed numbers add up to a whole number. For example, look at $6\frac{1}{7} + 12\frac{6}{7}$. Add the whole number parts of each number: $6 + 12 = 18$. Add the fractions: $\frac{1}{7} + \frac{6}{7} = \frac{7}{7}$, which is equal to 1. Add that to the sum of the whole numbers: $18 + 1 = 19$. Therefore, $6\frac{1}{7} + 12\frac{6}{7} = 19$.

The denominators of the fractions in mixed numbers can be unlike, just as when we looked at adding fractions in the last chapter. And just as in the last chapter, to add mixed numbers with unlike denominators, we must find common denominators before adding.

Example

$$9\frac{2}{3} + 15\frac{7}{15}$$

First, add the whole numbers: $9 + 15 = 24$. Next, add the fractions. The least common denominator of 3 and 15 is 15: $\frac{2}{3} = \frac{10}{15}$. Now, we can add the fractions: $\frac{10}{15} + \frac{7}{15} = \frac{17}{15}$. Because 17 divided by 15 is 1 with 2 left over, $\frac{17}{15} = 1\frac{2}{15}$. Finally, add this mixed number to the sum of the whole numbers: $24 + 1\frac{2}{15} = 25\frac{2}{15}$.

DIRECTIONS: Add the mixed numbers and simplify your answers. You can check your answers at the end of this chapter.

11. $4\frac{4}{9} + 11\frac{1}{9}$

12. $13\frac{2}{3} + 5\frac{2}{3}$

13. $15\frac{6}{8} + 8\frac{2}{8}$

14. $9\frac{7}{11} + 10\frac{15}{11}$

15. $1\frac{3}{8} + \frac{3}{4}$

16. $7\frac{5}{6} + 7\frac{4}{5}$

17. $10\frac{3}{7} + 3\frac{7}{10}$

18. $20\frac{1}{5} + 17\frac{5}{8}$

INSIDE TRACK

THERE IS ANOTHER way to add mixed numbers. Rather than adding the whole number parts and the fraction parts separately, and then combining them, we can instead begin by converting both mixed numbers to improper fractions. Then, the problem becomes adding fractions instead of adding mixed numbers. By now, we have plenty of experience adding fractions! Here's an example:

$$2\frac{2}{5} + 3\frac{1}{4}$$

Convert each mixed number to an improper fraction. $2\frac{2}{5} = \frac{12}{5}$ and $3\frac{1}{4} = \frac{13}{4}$. The problem is now $\frac{12}{5} + \frac{13}{4}$. Because 20 is the least common multiple of 5 and 4, convert both fractions to a number over 20: $\frac{12}{5} = \frac{48}{20}$ and $\frac{13}{4} = \frac{65}{20}$. $\frac{48}{20} + \frac{65}{20} = \frac{113}{20}$. Finally, we convert our answer to a mixed number; 113 divided by 20 is 5 with 13 left over, so $\frac{113}{20} = 5\frac{13}{20}$.

SUBTRACTING MIXED NUMBERS

We can use a similar strategy to subtract mixed numbers. In a subtraction problem involving two mixed numbers, subtract the whole number part of the second mixed number from the whole number part of the first mixed number. Then, subtract the fraction part of the second mixed number from the fraction part of the first mixed number.

Example

$$9\frac{7}{9} - 2\frac{2}{9}$$

Work with the whole numbers first: $9 - 2 = 7$. Next, work with the fractions. $\frac{7}{9} - \frac{2}{9} = \frac{5}{9}$. The difference is a simplified, proper fraction, so our answer is $7\frac{5}{9}$.

We've seen how the sum of two mixed numbers can be just a whole number. The same goes for differences.

Example

$$5\frac{1}{3} - 3\frac{1}{3}$$

Work with the whole numbers first: $5 - 3 = 2$. Next, work with the fractions. $\frac{1}{3} - \frac{1}{3} = 0$. Our answer is simply the whole number 2.

As you might have guessed, the difference between two mixed numbers might not contain a whole number: $4\frac{7}{8} - 4\frac{2}{8}$ is equal to $\frac{5}{8}$, because $4 - 4 = 0$.

Just as you need common denominators to add mixed numbers, you need common denominators to subtract mixed numbers.

Example

$$10\frac{9}{10} - 2\frac{3}{4}$$

Work with the whole numbers first: $10 - 2 = 8$. Next, work with the fractions. The least common multiple of 10 and 4 is 40. $\frac{9}{10} = \frac{36}{40}$ and $\frac{3}{4} = \frac{30}{40}$. Now we can subtract: $\frac{36}{40} - \frac{30}{40} = \frac{6}{40}$. The greatest common factor of 6 and 40 is 2, so $\frac{6}{40}$ reduces to $\frac{3}{20}$. Our answer is $8\frac{3}{20}$.

Subtracting mixed numbers can get a little tricky when the fraction part of the mixed number that is being subtracted is greater than the fraction part of the mixed number from which it is being subtracted. When that happens, we must borrow from the whole number part of our answer.

Example

$$5\frac{1}{3} - 1\frac{2}{3}$$

Work with the whole numbers first: $5 - 1 = 4$. Now look at the fractions. We can't subtract $\frac{2}{3}$ from $\frac{1}{3}$, because $\frac{2}{3}$ is greater than $\frac{1}{3}$. We need to borrow 1 from the whole number part of our answer: $4 - 1 = 3$. The whole number part of our answer is now 3. What do we do with the 1 we just borrowed? We write it as $\frac{3}{3}$ and add it to the fraction that we already have, $\frac{1}{3}$: $\frac{3}{3} + \frac{1}{3} = \frac{4}{3}$. Now we can subtract the fractions: $\frac{4}{3} - \frac{2}{3} = \frac{2}{3}$. The fractional part of our answer is $\frac{2}{3}$. Therefore, $5\frac{1}{3} - 1\frac{2}{3} = 3\frac{2}{3}$.

Instead of subtracting $1\frac{2}{3}$ from $5\frac{1}{3}$, we subtracted $1\frac{2}{3}$ from $4\frac{4}{3}$, because $5\frac{1}{3}$ and $4\frac{4}{3}$ are equal. When we borrow 1 from the whole number part of the mixed number, we write that 1 as a number over the denominator of the mixed number. Because the denominator of the mixed number $5\frac{1}{3}$ was 3, we wrote the borrowed 1 as $\frac{3}{3}$.

Let's look at another example just to be sure we've got it down!

Example

$$15\frac{2}{7} - 12\frac{7}{8}$$

Again, work with the whole numbers first: $15 - 12 = 3$. Now look at the fractions. The least common denominator of 7 and 8 is 56; $\frac{2}{7} = \frac{16}{56}$ and $\frac{7}{8} = \frac{49}{56}$. We can't subtract $\frac{49}{56}$ from $\frac{16}{56}$, because $\frac{49}{56}$ is greater than $\frac{16}{56}$. We need to borrow 1 from the whole number part of our answer: $3 - 1 = 2$. We write the borrowed 1 as $\frac{56}{56}$ and add it to the fraction that we already have: $\frac{16}{56} + \frac{56}{56} = \frac{72}{56}$. Now we can subtract the fractions: $\frac{72}{56} - \frac{49}{56} = \frac{23}{56}$. The fractional part of our answer is $\frac{23}{56}$. Therefore, $15\frac{2}{7} - 12\frac{7}{8} = 2\frac{23}{56}$.

CAUTION!

WHEN YOU BORROW, be sure to subtract 1 from the whole number part of your answer, and be sure that you add that borrowed 1 to the correct fraction. In a subtraction sentence, the number that you are subtracting is called the **subtrahend**, and the number from which you are subtracting is called the **minuend**. In the subtraction sentence $5 - 4 = 1$, 5 is the minuend and 4 is the subtrahend. When you borrow, be sure to add 1 to the fraction part of the minuend. In the last example we saw, $15\frac{2}{7}$ was the minuend. After borrowing 1 from the whole number part of our answer, we added 1 in the form of $\frac{56}{56}$ to $\frac{16}{56}$, the fractional part of the minuend.

Some people prefer to convert both mixed numbers to improper fractions before subtracting. If you do this, you won't have to worry about borrowing at all. Converting mixed numbers to improper fractions is good practice, and as we'll soon see, it's the only way to multiply and divide mixed numbers. Let's look at a subtraction example first.

Example

$$3\frac{1}{2} - 2\frac{4}{5}$$

First, convert each number to an improper fraction. $3 \times 2 = 6$, $6 + 1 = 7$, so $3\frac{1}{2} = \frac{7}{2}$. $2 \times 5 = 10$, $10 + 4 = 14$, so $2\frac{4}{5} = \frac{14}{5}$. The least common denominator of 2 and 5 is 10; $\frac{7}{2} = \frac{35}{10}$ and $\frac{14}{5} = \frac{28}{10}$. Now we can subtract: $\frac{35}{10} - \frac{28}{10} = \frac{7}{10}$.

INSIDE TRACK

IF YOU CAN see that a subtraction problem involving mixed numbers will NOT require borrowing, the easiest method is to subtract whole numbers and subtract fractions. If you can see that the problem will require borrowing, the easiest method is to convert each mixed number to an improper fraction before subtracting. However, either method will work in either case. If you find one method easier than the other, use it all the time!

PRACTICE LAP

DIRECTIONS: Subtract the mixed numbers and simplify your answers. You can check your answers at the end of this chapter.

19. $9\frac{3}{4} - 6\frac{1}{4}$

20. $16\frac{5}{6} - 2\frac{1}{6}$

21. $7\frac{4}{7} - 3\frac{5}{7}$

22. $17\frac{1}{5} - 6\frac{4}{5}$

23. $8\frac{2}{3} - 5\frac{1}{2}$

24. $11\frac{7}{12} - 4\frac{3}{8}$

25. $3\frac{2}{9} - 1\frac{5}{6}$

26. $12\frac{3}{4} - 10\frac{6}{7}$

MULTIPLYING MIXED NUMBERS

There is only one method for multiplying mixed numbers: Convert each mixed number to an improper fraction, and then multiply the numerators and multiply the denominators. No common denominators needed—once we have two improper fractions, we're ready to go!

Example

$$2\tfrac{1}{6} \times 6\tfrac{2}{3}$$

Convert each number to an improper fraction. $2 \times 6 = 12$, $12 + 1 = 13$, so $2\tfrac{1}{6} = \tfrac{13}{6}$; $6 \times 3 = 18$, $18 + 2 = 20$, so $6\tfrac{2}{3} = \tfrac{20}{3}$. Now that we have two improper fractions, multiply the numerators and multiply the denominators: $13 \times 20 = 260$ and $6 \times 3 = 18$, so $\tfrac{13}{6} \times \tfrac{20}{3} = \tfrac{260}{18}$. The greatest common factor of 260 and 18 is 2, so $\tfrac{260}{18}$ reduces to $\tfrac{130}{9}$. 130 divided by 9 is 14 with 4 left over, so $\tfrac{130}{9} = 14\tfrac{4}{9}$.

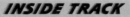

INSIDE TRACK

THERE ARE THREE places in which you can simplify a multiplication problem involving mixed numbers. After converting the mixed numbers to fractions, you can divide a numerator and denominator by the same number. In the last example, we could have simplified $\tfrac{13}{6} \times \tfrac{20}{3}$ by dividing the 6 in the first fraction and the 20 in the second fraction by 2. Or, we could have reduced $\tfrac{260}{18}$ to $\tfrac{130}{9}$ before converting to a mixed number (which is what we did). Finally, we can reduce the fraction part of a mixed number after converting the improper fraction into a mixed number. You can simplify at any—or all three—places. Often, if you reduce the improper fractions first, you'll have an easier time multiplying—and you may not have to simplify at all later.

Let's look at one more example.

Example

$$8\frac{4}{7} \times 3\frac{3}{5}$$

Convert each number to an improper fraction: $8 \times 7 = 56$, and $56 + 4 = 60$, so $8\frac{4}{7} = \frac{60}{7}$; $3 \times 5 = 15$, and $15 + 3 = 18$, so $3\frac{3}{5} = \frac{18}{5}$. Now we have the multiplication problem $\frac{60}{7} \times \frac{18}{5}$. Divide the 60 in the first fraction and the 5 in the second fraction by 5. The problem becomes $\frac{12}{7} \times 18$; $12 \times 18 = 216$; 216 divided by 7 is 30 with 6 left over, so $8\frac{4}{7} \times 3\frac{3}{5} = 30\frac{6}{7}$.

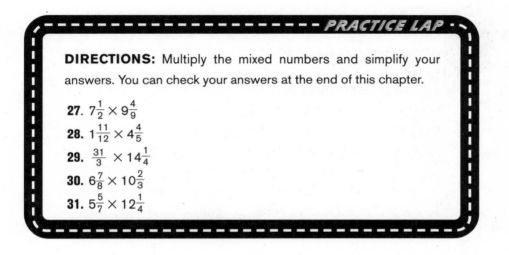

PRACTICE LAP

DIRECTIONS: Multiply the mixed numbers and simplify your answers. You can check your answers at the end of this chapter.

27. $7\frac{1}{2} \times 9\frac{4}{9}$

28. $1\frac{11}{12} \times 4\frac{4}{5}$

29. $\frac{31}{3} \times 14\frac{1}{4}$

30. $6\frac{7}{8} \times 10\frac{2}{3}$

31. $5\frac{5}{7} \times 12\frac{1}{4}$

PACE YOURSELF

A PIZZERIA NEEDS to cater 6 different parties, and each party needs $2\frac{1}{2}$ pizzas. How many pizzas does the pizzeria need to make? What if the pizzeria was catering 9 parties? What if each party needed $3\frac{1}{4}$ pizzas? Besides this example of a pizzeria, what other real-life situations can you think of that would require multiplying mixed numbers?

DIVIDING MIXED NUMBERS

You've probably noticed the similarities between adding, subtracting, multiplying, and dividing fractions and adding, subtracting, multiplying, and dividing mixed numbers. You might have already guessed how to divide mixed numbers: Convert them to improper fractions, find the reciprocal of the divisor, and multiply. No common denominators (and no division!) needed.

CAUTION!

BECAUSE MIXED NUMBERS must be converted to improper fractions before dividing, it would be easy to forget to take the reciprocal of the second fraction (the divisor) before dividing. Don't let that happen to you! Start a division problem with mixed numbers by converting the divisor to an improper fraction FIRST—and then find its reciprocal right away. Then, convert the first fraction (the dividend) to an improper fraction. Now you're ready to multiply.

Example

$$1\frac{6}{8} \div 4\frac{7}{10}$$

Let's use the tip we just learned: convert the divisor, $4\frac{7}{10}$, to an improper fraction, and then find its reciprocal. $4 \times 10 = 40$, $40 + 7 = 47$, so $4\frac{7}{10} = \frac{47}{10}$. Remember, to find the reciprocal of a fraction, switch the numerator and the denominator. The reciprocal of $\frac{47}{10}$ is $\frac{10}{47}$. Now convert the dividend, $1\frac{6}{8}$, to an improper fraction. $1 \times 8 = 8$, and $8 + 6 = 14$, so $1\frac{6}{8} = \frac{14}{8}$. Our problem has become $\frac{14}{8} \times \frac{10}{47}$. Divide the 8 in the first fraction and the 10 in the second fraction by 2, and the problem becomes $\frac{14}{4} \times \frac{5}{47}$. We can also divide the 14 in the first fraction and the 4 in the first fraction by 2, making the problem $\frac{7}{2} \times \frac{5}{47}$. Multiply the numerators and multiply the denominators: $7 \times 5 = 35$ and $2 \times 47 = 94$, making our answer $\frac{35}{94}$.

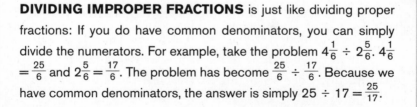

INSIDE TRACK

DIVIDING IMPROPER FRACTIONS is just like dividing proper fractions: If you do have common denominators, you can simply divide the numerators. For example, take the problem $4\frac{1}{6} \div 2\frac{5}{6}$. $4\frac{1}{6} = \frac{25}{6}$ and $2\frac{5}{6} = \frac{17}{6}$. The problem has become $\frac{25}{6} \div \frac{17}{6}$. Because we have common denominators, the answer is simply $25 \div 17 = \frac{25}{17}$.

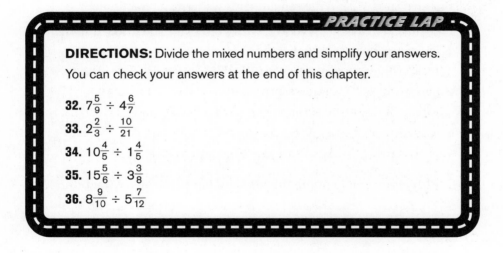

PRACTICE LAP

DIRECTIONS: Divide the mixed numbers and simplify your answers. You can check your answers at the end of this chapter.

32. $7\frac{5}{9} \div 4\frac{6}{7}$

33. $2\frac{2}{3} \div \frac{10}{21}$

34. $10\frac{4}{5} \div 1\frac{4}{5}$

35. $15\frac{5}{6} \div 3\frac{3}{8}$

36. $8\frac{9}{10} \div 5\frac{7}{12}$

By first learning how to add, subtract, multiply, and divide fractions, we established a solid foundation for learning how to add, subtract, multiply, and divide mixed numbers. After all, performing those operations on mixed numbers was just like working with fractions—just really large, improper fractions. We've got a few topics on fractions left to cover, and then it's on to decimals.

ANSWERS

1. Thirteen divided by 2 is 6, with 1 left over. The whole number part of our answer is 6, and the remainder is written as a fraction over 2: $6\frac{1}{2}$.

2. Twenty-two divided by 6 is 3, with 4 left over. The whole number part of our answer is 3, and the remainder is written as a fraction over 6: $3\frac{4}{6}$. The fraction $\frac{4}{6}$ can be simplified. The greatest common factor of 4 and 6 is 2, and because $\frac{4}{2} = 2$ and $\frac{6}{2} = 3$, $\frac{4}{6} = \frac{2}{3}$. $\frac{22}{6} = 3\frac{2}{3}$.

3. Fifty divided by 9 is 5, with 5 left over. The whole number part of our answer is 5, and the remainder is written as a fraction over 9: $5\frac{5}{9}$.

4. Eighty-one divided by 7 is 11, with 4 left over. The whole number part of our answer is 11, and the remainder is written as a fraction over 7: $11\frac{4}{7}$.

5. One hundred divided by 8 is 12, with 4 left over. The whole number part of our answer is 12, and the remainder is written as a fraction over 8: $12\frac{4}{8}$. The fraction $\frac{4}{8}$ can be simplified. The greatest common factor of 4 and 8 is 4, and because $\frac{4}{4} = 1$ and $\frac{8}{4} = 2$, $\frac{4}{8} = \frac{1}{2}$. $\frac{100}{8} = 12\frac{1}{2}$.

6. Multiply the whole number by the denominator of the fraction: $5 \times 5 = 25$. Add the numerator of the fraction to that product: $25 + 4 = 29$. Write that sum over the denominator of the fraction: $\frac{29}{5}$.

7. Multiply the whole number by the denominator of the fraction: $4 \times 7 = 28$. Add the numerator of the fraction to that product: $28 + 2 = 30$. Write that sum over the denominator of the fraction: $\frac{30}{7}$.

8. Multiply the whole number by the denominator of the fraction: $1 \times 10 = 10$. Add the numerator of the fraction to that product: $10 + 3 = 13$. Write that sum over the denominator of the fraction: $\frac{13}{10}$.

9. Multiply the whole number by the denominator of the fraction: $10 \times 8 = 80$. Add the numerator of the fraction to that product: $80 + 7 = 87$. Write that sum over the denominator of the fraction: $\frac{87}{8}$.

10. Multiply the whole number by the denominator of the fraction: $3 \times 12 = 36$. Add the numerator of the fraction to that product: $36 + 5 = 41$. Write that sum over the denominator of the fraction: $\frac{41}{12}$.

11. Add the whole numbers: $4 + 11 = 15$. Add the fractions: $\frac{4}{9} + \frac{1}{9} = \frac{5}{9}$. $4\frac{4}{9} + 11\frac{1}{9} = 15\frac{5}{9}$.

12. Add the whole numbers: $13 + 5 = 18$. Add the fractions: $\frac{2}{3} + \frac{2}{3} = \frac{4}{3}$. Convert $\frac{4}{3}$ to a mixed number: $\frac{4}{3} = 1\frac{1}{3}$. Add it to the sum of the whole numbers: $18 + 1\frac{1}{3} = 19\frac{1}{3}$.

13. Add the whole numbers: $15 + 8 = 23$. Add the fractions: $\frac{6}{8} + \frac{2}{8} = \frac{8}{8}$. Convert $\frac{8}{8}$ to a whole number: $\frac{8}{8} = 1$. Add it to the sum of the whole numbers: $23 + 1 = 24$.

14. Add the whole numbers: $9 + 10 = 19$. Add the fractions: $\frac{7}{11} + \frac{15}{11} = \frac{22}{11}$. Convert $\frac{22}{11}$ to a whole number: $\frac{22}{11} = 2$. Add it to the sum of the whole numbers: $19 + 2 = 21$.

15. Add the whole numbers: $1 + 0 = 1$. To add the fractions, find common denominators. The least common denominator of 8 and 4 is 8: $\frac{3}{4} = \frac{6}{8}$. $\frac{3}{8} + \frac{6}{8} = \frac{9}{8}$. Convert $\frac{9}{8}$ to a mixed number: $\frac{9}{8} = 1\frac{1}{8}$. Add it to the sum of the whole numbers: $1 + 1\frac{1}{8} = 2\frac{1}{8}$.

16. Add the whole numbers: $7 + 7 = 14$. To add the fractions, find common denominators. The least common denominator of 6 and 5 is 30: $\frac{5}{6} = \frac{25}{30}$ and $\frac{4}{5} = \frac{24}{30}$. $\frac{25}{30} + \frac{24}{30} = \frac{49}{30}$. Convert $\frac{49}{30}$ to a mixed number: $\frac{49}{30} = 1\frac{19}{30}$. Add it to the sum of the whole numbers: $14 + 1\frac{19}{30} = 15\frac{19}{30}$.

17. Add the whole numbers: $10 + 3 = 13$. To add the fractions, find common denominators. The least common denominator of 7 and 10 is 70. $\frac{3}{7} = \frac{30}{70}$ and $\frac{7}{10} = \frac{49}{70}$. $\frac{30}{70} + \frac{49}{70} = \frac{79}{70}$. Convert $\frac{79}{70}$ to a mixed number: $\frac{79}{70} = 1\frac{9}{70}$. Add it to the sum of the whole numbers: $13 + 1\frac{9}{70} = 14\frac{9}{70}$.

18. Add the whole numbers: $20 + 17 = 37$. To add the fractions, find common denominators. The least common denominator of 5 and 8 is 40. $\frac{1}{5} = \frac{8}{40}$ and $\frac{5}{8} = \frac{25}{40}$. $\frac{8}{40} + \frac{25}{40} = \frac{33}{40}$. $20\frac{1}{5} + 17\frac{5}{8} = 37\frac{33}{40}$.

19. Subtract the whole numbers: $9 - 6 = 3$. Subtract the fractions: $\frac{3}{4} - \frac{1}{4} = \frac{2}{4}$, which reduces to $\frac{1}{2}$. $9\frac{3}{4} - 6\frac{1}{4} = 3\frac{1}{2}$.

20. Subtract the whole numbers: $16 - 2 = 14$. Subtract the fractions: $\frac{5}{6} - \frac{1}{6} = \frac{4}{6}$, which reduces to $\frac{2}{3}$. $16\frac{5}{6} - 2\frac{1}{6} = 14\frac{2}{3}$.

21. Convert both mixed numbers to improper fractions. $7\frac{4}{7} = \frac{53}{7}$ and $3\frac{5}{7} = \frac{26}{7}$. $\frac{53}{7} - \frac{26}{7} = \frac{27}{7}$. Convert the improper fraction back to a mixed number: 27 divided by 7 is 3 with 6 left over, so $\frac{27}{7} = 3\frac{6}{7}$.

22. Convert both mixed numbers to improper fractions. $17\frac{1}{5} = \frac{86}{5}$ and $6\frac{4}{5} = \frac{34}{5}$. $\frac{86}{5} - \frac{34}{5} = \frac{52}{5}$. Convert the improper fraction back to a mixed number. 52 divided by 5 is 10 with 2 left over, so $\frac{52}{5} = 10\frac{2}{5}$.

23. Subtract the whole numbers: $8 - 5 = 3$. The least common multiple of 3 and 2 is 6, so convert both fractions to a number over 6. $\frac{2}{3} = \frac{4}{6}$ and $\frac{1}{2} = \frac{3}{6}$. Subtract the fractions: $\frac{4}{6} - \frac{3}{6} = \frac{1}{6}$. $8\frac{2}{3} - 5\frac{1}{2} = 3\frac{1}{6}$.

24. Subtract the whole numbers: $11 - 4 = 7$. The least common multiple of 12 and 8 is 24, so convert both fractions to a number over 24. $\frac{7}{12} = \frac{14}{24}$ and $\frac{3}{8} = \frac{9}{24}$. Subtract the fractions: $\frac{14}{24} - \frac{9}{24} = \frac{5}{24}$. $11\frac{7}{12} - 4\frac{3}{8} = 7\frac{5}{24}$.

25. Convert both mixed numbers to improper fractions. $3\frac{2}{9} = \frac{29}{9}$ and $1\frac{5}{6} = \frac{11}{6}$. The least common multiple of 9 and 6 is 54, so convert both fractions to a number over 54. $\frac{29}{9} = \frac{174}{54}$ and $\frac{11}{6} = \frac{99}{54}$. Subtract the fractions: $\frac{174}{54} - \frac{99}{54} = \frac{75}{54}$. Convert the improper fraction back to a mixed number: 75 divided by 54 is 1 with 21 left over, so $\frac{75}{54} = 1\frac{21}{54}$.

26. Convert both mixed numbers to improper fractions. $12\frac{3}{4} = \frac{51}{4}$ and $10\frac{6}{7} = \frac{76}{7}$. The least common multiple of 4 and 7 is 28, so convert both fractions to a number over 28. $\frac{51}{4} = \frac{357}{28}$ and $\frac{76}{7} = \frac{304}{28}$. Subtract the fractions: $\frac{357}{28} - \frac{304}{28} = \frac{53}{28}$. Convert the improper fraction back to a mixed number. 53 divided by 28 is 1 with 25 left over, so $\frac{53}{28} = 1\frac{25}{28}$.

27. Convert both mixed numbers to improper fractions. $7\frac{1}{2} = \frac{15}{2}$ and $9\frac{4}{9} = \frac{85}{9}$. Divide the 15 in the first fraction and the 9 in the second fraction by 3. The problem becomes $\frac{5}{2} \times \frac{85}{3}$. $5 \times 85 = 425$ and $2 \times 3 = 6$. 425 divided by 6 is 70 with 5 left over: $70\frac{5}{6}$.

28. Convert both mixed numbers to improper fractions. $1\frac{11}{12} = \frac{23}{12}$ and $4\frac{4}{5} = \frac{24}{5}$. Divide the 12 in the first fraction and the 24 in the second fraction by 2. The problem becomes $23 \times \frac{2}{5}$. $23 \times 2 = 46$. 46 divided by 5 is 9 with 1 left over: $9\frac{1}{5}$.

29. Convert both mixed numbers to improper fractions. $3\frac{1}{3} = \frac{10}{3}$ and $14\frac{1}{4} = \frac{57}{4}$. Divide the 3 in the first fraction and the 57 in the second fraction by 3, and divide the 10 in the first fraction and the 4 in the second fraction by 2. The problem becomes $5 \times \frac{19}{2}$. $5 \times 19 = 95$. 95 divided by 2 is 47 with 1 left over: $47\frac{1}{2}$.

30. Convert both mixed numbers to improper fractions. $6\frac{7}{8} = \frac{55}{8}$ and $10\frac{2}{3} = \frac{32}{3}$. Divide the 8 in the first fraction and the 32 in the second fraction by 8. The problem becomes $55 \times \frac{4}{3}$. $55 \times 4 = 220$. 220 divided by 3 is 73 with 1 left over: $73\frac{1}{3}$.

31. Convert both mixed numbers to improper fractions. $5\frac{5}{7} = \frac{40}{7}$ and $12\frac{1}{4} = \frac{49}{4}$. Divide the 7 in the first fraction and the 49 in the second fraction by 7, and divide the 40 in the first fraction and the 4 in the second fraction by 4. The problem becomes $10 \times 7 = 70$.

32. Convert both mixed numbers to improper fractions. $7\frac{5}{9} = \frac{68}{9}$ and $4\frac{6}{7} = \frac{34}{7}$. The reciprocal of the divisor, $\frac{34}{7}$, is $\frac{7}{34}$. The problem is now $\frac{68}{9} \times \frac{7}{34}$. Divide the 68 in the first fraction and the 34 in the second fraction by 34. The problem is now $\frac{2}{9} \times 7 = \frac{14}{9}$. 14 divided by 9 is 1 with 5 left over, so $\frac{14}{9} = 1\frac{5}{9}$.

33. Convert the mixed number to an improper fraction. $2\frac{2}{3} = \frac{8}{3}$. The reciprocal of the divisor, $\frac{10}{21}$, is $\frac{21}{10}$. The problem is now $\frac{8}{3} \times \frac{21}{10}$. Divide the 8 in the first fraction and the 10 in the second fraction by 2, and divide the 3 in the first fraction and the 21 in the second fraction by 3. The problem is now $4 \times \frac{7}{5} = \frac{28}{5}$. 28 divided by 5 is 5 with 3 left over, so $\frac{28}{5} = 5\frac{3}{5}$.

34. Convert both mixed numbers to improper fractions. $10\frac{4}{5} = \frac{54}{5}$ and $1\frac{4}{5} = \frac{9}{5}$. Because we have common denominators, we can divide the numerators: 54 divided by 9 is 6.

35. Convert both mixed numbers to improper fractions. $15\frac{5}{6} = \frac{95}{6}$ and $3\frac{3}{8} = \frac{27}{8}$. The reciprocal of the divisor, $\frac{27}{8}$, is $\frac{8}{27}$. The problem is now $\frac{95}{6} \times \frac{8}{27}$. Divide the 6 in the first fraction and the 8 in the second fraction by 2. The problem is now $\frac{95}{3} \times \frac{4}{27} = \frac{380}{81}$. 380 divided by 81 is 4 with 56 left over, so $\frac{380}{81} = 4\frac{56}{81}$.

36. Convert both mixed numbers to improper fractions. $8\frac{9}{10} = \frac{89}{10}$ and $5\frac{7}{12} = \frac{67}{12}$. The reciprocal of the divisor, $\frac{67}{12}$, is $\frac{12}{67}$. The problem is now $\frac{89}{10} \times \frac{12}{67}$. Divide the 10 in the first fraction and the 12 in the second fraction by 2. The problem is now $\frac{89}{5} \times \frac{6}{67} = \frac{534}{335}$. 534 divided by 335 is 1 with 199 left over, so $\frac{534}{335} = 1\frac{199}{335}$.

Fraction Leftovers

WHAT'S AROUND THE BEND

➡ Complex Fractions
➡ Ratios
➡ Proportions

We've covered almost everything there is to know
about fractions—almost. In this chapter, we'll look at two more
ways to use fractions (as ratios and proportions) and one other type of fraction: complex fractions.

COMPLEX FRACTIONS

A proper fraction has a value that is between –1 and 1. Therefore, the numerator of a proper fraction is less than the denominator of the fraction. For
instance, in the proper fraction $\frac{1}{5}$, the numerator, 1, is less than the denominator, 5. In improper fractions, the numerator is greater than or equal to the
denominator of the fraction. For instance, in the improper fraction $\frac{8}{3}$, the
numerator, 8, is greater than the denominator, 3. In both of these fractions,
the numerators and denominators are whole numbers. But what if they were
fractions?

FUEL FOR THOUGHT

A COMPLEX FRACTION is a fraction whose numerator or denominator (or both) is a fraction. For instance, $\frac{(\frac{4}{5})}{3}$ is a complex fraction, as is $\frac{1}{(\frac{6}{7})}$. In fact, so is $\frac{(\frac{3}{4})}{(\frac{2}{9})}$. The fractions in a complex fraction can be proper or improper. The numerator or denominator of a complex fraction could be a complex fraction itself.

At first glance, it seems like complex fractions would be hard to add, subtract, multiply, and divide. However, as with mixed numbers, we can simplify them to proper or improper fractions, and we already know how to work with those. So how do you simplify a complex fraction?

Remember, the fraction bar in a fraction represents division. A simple fraction, such as $\frac{7}{10}$, means "7 divided by 10." A complex fraction reads the same way. $\frac{\frac{1}{2}}{8}$ means "$\frac{1}{2}$ divided by 8," or $\frac{1}{2} \div 8$. Any whole number can be rewritten as a fraction with a denominator 1, so $\frac{1}{2} \div 8$ is the same as $\frac{1}{2} \div \frac{8}{1}$. Now the complex fraction has become a division problem that we know how to solve. The reciprocal of $\frac{8}{1}$ is $\frac{1}{8}$. $\frac{1}{2} \times \frac{1}{8} = \frac{1}{16}$. The complex fraction $\frac{(\frac{1}{2})}{8}$ is equivalent to the simple fraction $\frac{1}{16}$.

Example

$$\frac{5}{\left(\frac{3}{4}\right)}$$

To simplify a complex fraction, begin by replacing the fraction bar with the division symbol. $\frac{5}{\left(\frac{3}{4}\right)} = 5 \div \frac{3}{4}$. If one part of the complex fraction is a whole number, rewrite it as a fraction with a denominator of 1: $\frac{5}{1} \div \frac{3}{4}$. Now, divide: $\frac{5}{1} \div \frac{3}{4} = \frac{5}{1} \times \frac{4}{3} = \frac{20}{3} = 6\frac{2}{3}$.

INSIDE TRACK

WHEN A COMPLEX fraction has a whole number in the numerator and a proper fraction in the denominator, when simplified, the fraction will have a value that is greater than the numerator of the complex fraction, as in the last example. Why? When a whole number is divided by a number that is between 0 and 1, it becomes larger.

When a complex fraction has a whole number in the denominator and a proper fraction in the numerator, when simplified, the fraction will have a value that is less than the numerator of the complex fraction. When a fraction is divided by a whole number, it becomes smaller.

Sometimes, a complex fraction has fractions in both the numerator and the denominator.

Example

$$\frac{\left(\frac{8}{9}\right)}{\left(\frac{5}{7}\right)}$$

Begin by replacing the fraction bar with the division symbol $\frac{\left(\frac{8}{9}\right)}{\left(\frac{5}{7}\right)}$. = $\frac{8}{9} \div \frac{5}{7}$. Just like that, the complex fraction is a division problem that, by now, we've seen many times: $\frac{8}{9} \div \frac{5}{7} = \frac{8}{9} \times \frac{7}{5} = \frac{56}{45} = 1\frac{11}{45}$. Our answer is a mixed number. When $\frac{8}{9}$ is divided by a number between 0 and 1 (a proper fraction), it becomes larger.

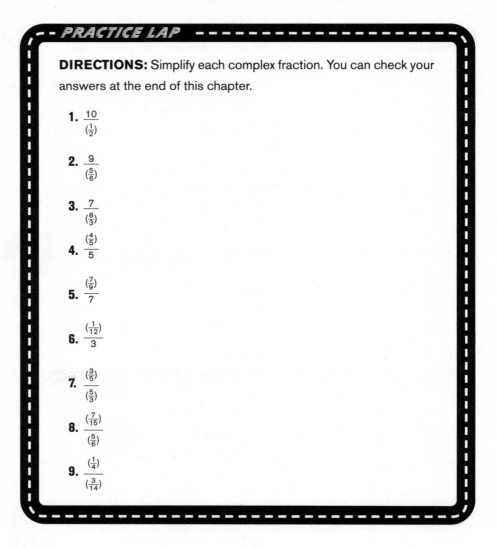

PRACTICE LAP

DIRECTIONS: Simplify each complex fraction. You can check your answers at the end of this chapter.

1. $\dfrac{10}{\left(\frac{1}{2}\right)}$

2. $\dfrac{9}{\left(\frac{5}{6}\right)}$

3. $\dfrac{7}{\left(\frac{8}{3}\right)}$

4. $\dfrac{\left(\frac{4}{5}\right)}{5}$

5. $\dfrac{\left(\frac{7}{9}\right)}{7}$

6. $\dfrac{\left(\frac{1}{12}\right)}{3}$

7. $\dfrac{\left(\frac{3}{5}\right)}{\left(\frac{5}{3}\right)}$

8. $\dfrac{\left(\frac{7}{15}\right)}{\left(\frac{5}{6}\right)}$

9. $\dfrac{\left(\frac{1}{4}\right)}{\left(\frac{3}{14}\right)}$

Because a complex fraction is just a division problem, we can turn division problems into complex fractions. Just as the division problem "6 divided by 9" is the fraction $\frac{6}{9}$, the division problem "$\frac{2}{3}$ divided by 7" is the complex fraction $\dfrac{\left(\frac{2}{3}\right)}{7}$. The dividend of a division problem becomes the numerator of the complex fraction and the divisor of the problem becomes the denominator of the complex fraction.

Example

$$\frac{1}{10} \div \frac{8}{9}$$

The dividend, $\frac{1}{10}$, becomes the numerator and the divisor, $\frac{8}{9}$, becomes the denominator: $\frac{\left(\frac{1}{10}\right)}{\left(\frac{8}{9}\right)}$.

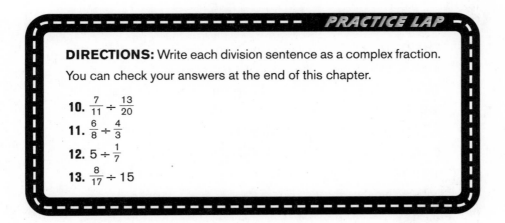

PRACTICE LAP

DIRECTIONS: Write each division sentence as a complex fraction. You can check your answers at the end of this chapter.

10. $\frac{7}{11} \div \frac{13}{20}$

11. $\frac{6}{8} \div \frac{4}{3}$

12. $5 \div \frac{1}{7}$

13. $\frac{8}{17} \div 15$

USING RATIOS TO COMPARE NUMBERS

So far, we've read fractions as division problems. But fractions can also represent ratios.

FUEL FOR THOUGHT

A RATIO IS a comparison, or relationship, between two numbers. Ratios are often shown with a colon, but can also be expressed as fractions. The ratio "2 to 1" can be written as 2:1 or $\frac{2}{1}$.

If the ratio of red marbles to blue marbles in a sack is 4 to 3, that ratio can be written as 4:3 or $\frac{4}{3}$. To convert a ratio in colon form to a fraction, make the first number of the ratio the numerator of the fraction and make the second number of the ratio the denominator of the fraction.

Example

A math class has 27 students. There are 15 boys and 12 girls in the class.

The ratio of boys to girls is 15:12. As a fraction, the ratio of boys to girls is $\frac{15}{12}$.

CAUTION!

THE ORDER OF the numbers in a ratio is very important. The ratio 4:3 is not the same as the ratio 3:4. Look again at the last example. The ratio of boys to girls is 15:12. If we had written the ratio as 12:15, that would be the ratio of *girls to boys*. Always match the numbers to their labels. Writing a ratio as a fraction can help you remember that the order of a ratio is important. The ratio of boys to girls is $\frac{15}{12}$. The fraction $\frac{15}{12}$ is not equal to the fraction $\frac{12}{15}$—just as the ratio 15:12 is not equal to the ratio 12:15.

Writing a ratio as a fraction helps remind us that ratios can be simplified. The fraction $\frac{15}{12}$ reduces to $\frac{5}{4}$. But wait a minute—there are 27 students in the class. Is the ratio 5:4, or $\frac{5}{4}$, still correct? Yes. A ratio represents a relationship between numbers, not an exact pair of numbers. This ratio says, "For every 5 boys, there are 4 girls." The following table shows the math class broken up into rows:

boy	girl	boy	girl	boy	girl
boy	girl	boy	girl	boy	girl
boy	girl	boy	girl	boy	girl
boy	girl	boy	girl	boy	girl
boy		boy		boy	

There are 3 rows of 5 boys and 3 rows of 4 girls. All 27 students are shown, and there are 5 boys for every 4 girls. The ratio holds true.

Example

A jar contains 11 lemon candies and 12 strawberry candies. The ratio of lemon candies to strawberry candies is 11:12 or $\frac{11}{12}$.

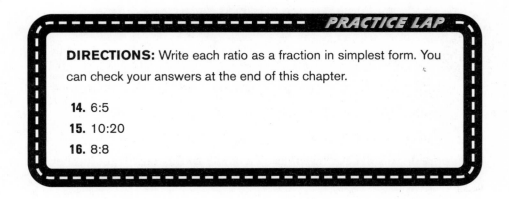

DIRECTIONS: Write each ratio as a fraction in simplest form. You can check your answers at the end of this chapter.

14. 6:5

15. 10:20

16. 8:8

RATIOS AS PROPORTIONS

We can use ratios to solve problems. Pairs of equivalent ratios, or fractions, are called proportions.

FUEL FOR THOUGHT

A RATIO IS a relationship between two numbers. A **proportion** is a relationship between two ratios. In fact, it's not just any relationship: a proportion is two equivalent ratios.

We've seen equivalent fractions before, like when we needed to find common denominators for two fractions. What is the fraction $\frac{1}{4}$ expressed as a number over 8? Because 8 is equal to 4 times 2, $\frac{1}{4} = \frac{2}{8}$. If $\frac{1}{4}$ and $\frac{2}{8}$ are ratios, then $\frac{1}{4} = \frac{2}{8}$ is a proportion.

A ratio is like a rule for a set. If we know the ratio of a set, we can create a proportion to help us find a missing piece of information. For instance, if we know the ratio of red shirts to blue shirts in a closet, and we know the total number of shirts in the closet, we can find the exact number of red shirts and blue shirts. Or, if we know the ratio and the exact number of red shirts, we can find the exact number of blue shirts.

Example

The ratio of red shirts to blue shirts in a closet is 2:3. If there are 4 red shirts in the closet, how many blue shirts are in the closet?

First, write the ratio as a fraction. $2:3 = \frac{2}{3}$. Next, set up a proportion to solve for the unknown value. In this ratio, the number of red shirts is the numerator and the number of blue shirts is the denominator. If there are 2 red shirts for every 3 blue shirts, then there are 4 red shirts for every x number of blue shirts. We use x to hold the place of the number of blue shirts, because that is the value we're trying to find. Now we can set up a proportion: $\frac{2}{3} = \frac{4}{x}$.

To solve a proportion for an unknown (the "x" in the proportion), we cross multiply. Multiply the numerator of the first fraction by the denominator of the second fraction, and multiply the denominator of the first fraction by the numerator of the second fraction. $(2)(x) = 2x$ and $(3)(4) = 12$. Then, set these products equal to each other: $2x = 12$. Solve for x by dividing both sides of the equation by 2: $\frac{2x}{2} = \frac{12}{2}$, $x = 6$.

What does $x = 6$ mean? We substitute 6 for x in our proportion: $\frac{2}{3} = \frac{4}{x}$, so $\frac{2}{3} = \frac{4}{6}$. $\frac{2}{3}$ is equivalent to $\frac{4}{6}$. If there are 2 red shirts for every 3 blue shirts in a closet, and there are 4 red shirts in the closet, then there must be 6 blue shirts in the closet.

What if there were 10 red shirts in the closet? Set up a new proportion and x will still represent the number of blue shirts in the closet. $\frac{2}{3} = \frac{10}{x}$. Cross multiply and set the products equal to each other: $(2)(x) = 2x$ and $(3)(10) = 30$. $2x = 30$. Again, divide both sides of the equation by 2 to find the value of x: $\frac{2x}{2} = \frac{30}{2}$, $x = 15$. If there are 10 red shirts in the closet, then there must be 15 blue shirts in the closet.

What if we knew the number of blue shirts in the closet, but not the number of red shirts? Let's say there are 18 blue shirts in the closet. How many red shirts are in the closet?

We must set up another proportion. The ratio of red shirts to blue shirts is still 2 to 3, but now the number of red shirts is unknown. $\frac{2}{3} = \frac{x}{18}$. Cross multiply and set the products equal to each other: $(2)(18) = (3)(x)$, $36 = 3x$. Divide both sides of the equation by 3: $\frac{36}{3} = \frac{3x}{3}$, $x = 12$. If there are 18 blue shirts in the closet, then there are 12 red shirts in the closet.

CAUTION!

BE SURE TO read ratios carefully. A fraction often means a part out of a whole or total, but that is not always the case with ratios. In the last example, the ratio $\frac{2}{3}$ represented a comparison between red shirts and blue shirts, and not "2 out of 3." It is not true that two-thirds of the shirts are red or that two-thirds of the shirts are blue. In the next section, we will see how ratios can represent a part out of a whole.

PRACTICE LAP

DIRECTIONS: Use proportions to solve each problem. You can check your answers at the end of this chapter.

17. The ratio of squares to triangles in a pattern is 3:4. If there are 24 squares in the pattern, how many triangles are in the pattern?

18. The ratio of pop albums to jazz albums in DeDe's collection is 6:5. If there are 30 jazz albums in her collection, how many pop albums are in her collection?

19. The ratio of dimes to pennies in Lindsay's bank is 3:10. If there are 70 pennies in her bank, how many dimes are in her bank?

20. The ratio of black cards to red cards in a deck is 7:2. If there are 28 black cards in the deck, how many red cards are in the deck?

We can also use proportions to find the number of a given item in a set if we know the total number of items in the set. If we know the ratio between the items in a set, we can rewrite that ratio as a part-to-whole relationship. Then, we can set that ratio equal to an unknown over the total.

Example

The ratio of fruit punch drinks to orange drinks is 5:9. If there are 56 total drinks, how many of them are fruit punch?

Because we have the ratio and the total, we must rewrite the ratio of fruit punch drinks to orange drinks as the ratio of fruit punch drinks to total drinks. Think about what the ratio 5:9 means: for every 5 fruit punch drinks, there are 9 orange drinks. In other words, for every 5 fruit punch drinks, there are $5 + 9 = 14$ total drinks. The ratio of fruit punch drinks to total drinks is 5:14, or $\frac{5}{14}$. We know the exact number of total drinks, 56, and we are looking for the exact number of fruit punch drinks. Use x to represent the exact number of fruit punch drinks. If 5 of every 14 total drinks are fruit punch, then x of 56 total drinks are fruit punch: $\frac{5}{14} = \frac{x}{56}$. Cross multiply and set the products equal to each other. $(5)(56) = (14)(x)$, $280 = 14x$. Divide both sides of the equation by 14: $\frac{280}{14} = \frac{14x}{14}$, $x = 20$. If there are 56 total drinks, then 20 of them are fruit punch.

But what if we wanted to know the number of orange drinks? If the ratio of fruit punch drinks to orange drinks is 5:9, then the ratio of orange drinks to total drinks is 9:14, or $\frac{9}{14}$. Set that ratio equal to x, the exact number of orange drinks, over 56, the exact number of total drinks: $\frac{9}{14} = \frac{x}{56}$. Cross multiply and set the products equal to each other. $(9)(56) = (14)(x)$, $504 = 14x$. Divide both sides of the equation by 14: $\frac{504}{14} = \frac{14x}{14}$, $x = 36$. If there are 56 total drinks, then 36 of them are orange.

IF YOU'VE JUST used a ratio to find the exact number of one type of item in a set, you can find the exact number of the other item—using subtraction. Let's look again at the last example. We used the ratio 5:14 and the total number of drinks, 56, to find the number of fruit punch drinks, 20. Rather than use a ratio to find the number of orange drinks, we can subtract the number of fruit punch drinks from the total number of drinks: 56 − 20 = 36 orange drinks, the same answer we found using ratios.

Example

The ratio of round cakes to sheet cakes in a bakery is 8:7. If there are 56 sheet cakes, how many total cakes are there?

Again, we will use the part-to-whole relationship to solve this problem. If there are 8 round cakes for every 7 sheet cakes, then there are 8 round cakes for every 7 + 8 = 15 total cakes, and there are 7 sheet cakes for every 15 total cakes. Because we know that there are exactly 56 sheet cakes, we need to use the ratio that compares the number of sheet cakes to the total number of cakes: 7:15, or $\frac{7}{15}$. If 7 out of every 15 cakes are sheet cakes, then 56 out of x total cakes are sheet cakes: $\frac{7}{15} = \frac{56}{x}$. Cross multiply and set the products equal to each other. $(7)(x) = (15)(56)$, $7x = 840$. Divide both sides of the equation by 7: $\frac{840}{7} = \frac{7x}{7}$, $x = 120$. If there are 56 sheet cakes, then there are 120 total cakes.

How many cakes are round cakes? Rather than use a ratio, we'll just subtract. If there are 120 total cakes, and 56 are sheet cakes, then 120 − 56 = 64 cakes are round cakes.

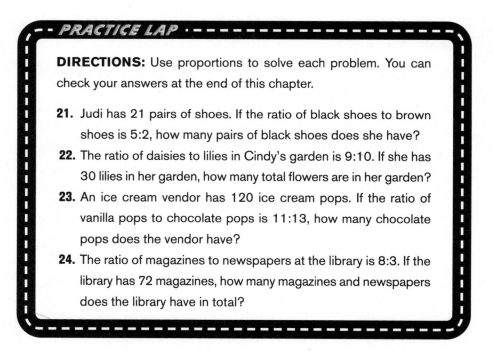

PACE YOURSELF

FIND THE RATIO of boys to girls in your math class. If there were 20 students in your class, and the ratio of boys to girls remained the same, how many boys would there be in your class? How many girls? Estimate the number of students in your school. If the ratio of boys to girls in your math class were the same as the ratio of boys to girls in your school, how many boys would there be in your school? How many girls?

PRACTICE LAP

DIRECTIONS: Use proportions to solve each problem. You can check your answers at the end of this chapter.

21. Judi has 21 pairs of shoes. If the ratio of black shoes to brown shoes is 5:2, how many pairs of black shoes does she have?

22. The ratio of daisies to lilies in Cindy's garden is 9:10. If she has 30 lilies in her garden, how many total flowers are in her garden?

23. An ice cream vendor has 120 ice cream pops. If the ratio of vanilla pops to chocolate pops is 11:13, how many chocolate pops does the vendor have?

24. The ratio of magazines to newspapers at the library is 8:3. If the library has 72 magazines, how many magazines and newspapers does the library have in total?

Complex fractions may not be so common, but ratios and proportions can be used to help us solve everyday problems. Now that you are a fractions expert, it's time to learn about decimals.

ANSWERS

1. Rewrite as a division sentence: $\dfrac{10}{\left(\frac{1}{2}\right)} = 10 \div \dfrac{1}{2} = 10 \times 2 = 20.$

2. Rewrite as a division sentence: $\dfrac{9}{\left(\frac{5}{6}\right)} = 9 \div \dfrac{5}{6} = 9 \times \dfrac{6}{5} = \dfrac{54}{5} = 10\dfrac{4}{5}.$

3. Rewrite as a division sentence: $\dfrac{7}{\left(\frac{8}{3}\right)} = 7 \div \dfrac{8}{3} = 7 \times \dfrac{3}{8} = \dfrac{21}{8} = 2\dfrac{5}{8}.$

4. Rewrite as a division sentence: $\dfrac{\left(\frac{4}{5}\right)}{5} = \dfrac{4}{5} \div 5 = \dfrac{4}{5} \times \dfrac{1}{5} = \dfrac{4}{25}.$

5. Rewrite as a division sentence: $\dfrac{\left(\frac{7}{9}\right)}{7} = \dfrac{7}{9} \div 7 = \dfrac{7}{9} \times \dfrac{1}{7} = \dfrac{1}{9} - \dfrac{1}{1} = \dfrac{1}{9}.$

6. Rewrite as a division sentence: $\dfrac{\left(\frac{1}{12}\right)}{3} = \dfrac{1}{12} \div 3 = \dfrac{1}{12} \times \dfrac{1}{3} = \dfrac{1}{36}.$

7. Rewrite as a division sentence: $\dfrac{\left(\frac{3}{5}\right)}{\left(\frac{5}{3}\right)} = \dfrac{3}{5} \div \dfrac{5}{3} = \dfrac{3}{5} \times \dfrac{3}{5} = \dfrac{9}{25}.$

8. Rewrite as a division sentence: $\dfrac{\left(\frac{7}{15}\right)}{\left(\frac{5}{6}\right)} = \dfrac{7}{15} \div \dfrac{5}{6} = \dfrac{7}{15} \times \dfrac{6}{5} = \dfrac{7}{5} \times \dfrac{2}{5} = \dfrac{14}{25}.$

9. Rewrite as a division sentence: $\dfrac{\left(\frac{1}{4}\right)}{\left(\frac{3}{4}\right)} = \dfrac{1}{4} \div \dfrac{3}{14} = \dfrac{1}{4} \times \dfrac{14}{3} = \dfrac{1}{2} \times \dfrac{7}{3} = \dfrac{7}{6} = 1\dfrac{1}{6}.$

10. $\dfrac{7}{11} \div \dfrac{13}{20} = \dfrac{\left(\frac{7}{11}\right)}{\left(\frac{13}{20}\right)}$

11. $\dfrac{6}{8} \div \dfrac{4}{3} = \dfrac{\left(\frac{6}{8}\right)}{\left(\frac{4}{3}\right)}$

12. $5 \div \dfrac{1}{7} = \dfrac{5}{\left(\frac{1}{7}\right)}$

13. $\dfrac{8}{17} \div 15 = \dfrac{\left(\frac{8}{17}\right)}{15}$

14. $6:5 = \dfrac{6}{5}$

15. $10:20 = \dfrac{10}{20} = \dfrac{1}{2}$

16. $8:8 = \dfrac{8}{8} = 1$

17. Write the ratio as a fraction: $3:4 = \dfrac{3}{4}$. Set up a proportion. If there are 3 squares for every 4 triangles, then there are 24 squares for x triangles: $\dfrac{3}{4} = \dfrac{24}{x}$. Cross multiply: $(3)(x) = (4)(24)$, $3x = 96$. Divide both sides

of the equation by 3: $\frac{3x}{3} = \frac{96}{3}$, $x = 32$. If there are 24 squares in the pattern, then there are 32 triangles in the pattern.

18. Write the ratio as a fraction: $6:5 = \frac{6}{5}$. Set up a proportion. If there are 6 pop albums for every 5 jazz albums, then there are x pop albums for 30 jazz albums: $\frac{6}{5} = \frac{x}{30}$. Cross multiply: $(6)(30) = (5)(x)$, $180 = 5x$. Divide both sides of the equation by 5: $\frac{180}{5} = \frac{5x}{5}$, $x = 36$. If there are 30 jazz albums in DeDe's collection, then there are 36 pop albums in DeDe's collection.

19. Write the ratio as a fraction: $3:10 = \frac{3}{10}$. Set up a proportion. If there are 3 dimes for every 10 pennies, then there are x dimes for 70 pennies: $\frac{3}{10} = \frac{x}{70}$. Cross multiply: $(3)(70) = (10)(x)$, $210 = 10x$. Divide both sides of the equation by 10: $\frac{210}{10} = \frac{10x}{10}$, $x = 21$. If there are 70 pennies in Lindsay's bank, then there are 21 dimes in her bank.

20. Write the ratio as a fraction: $7:2 = \frac{7}{2}$. Set up a proportion. If there are 7 black cards for every 2 red cards, then there are 28 black cards for x red cards: $\frac{7}{2} = \frac{28}{x}$. Cross multiply: $(7)(x) = (2)(28)$, $7x = 56$. Divide both sides of the equation by 7: $\frac{7x}{7} = \frac{56}{7}$, $x = 8$. If there are 28 black cards in the deck, then there are 8 red cards in the deck.

21. If the ratio of black shoes to brown shoes is 5:2, then the ratio of black shoes to total shoes is 5 to 5 + 2 = 7, or 5:7. Write the ratio as a fraction: $5:7 = \frac{5}{7}$. Set up a proportion. If 5 pairs of 7 total pairs are black, then x pairs of 21 total pairs are black: $\frac{5}{7} = \frac{x}{21}$. Cross multiply: $(7)(x) = (5)(21)$, $7x = 105$. Divide both sides of the equation by 7: $\frac{7x}{7} = \frac{105}{7}$, $x = 15$. If there are 21 total pairs of shoes, then 15 of them are black.

22. If the ratio of daisies to lilies is 9:10, then the ratio of lilies to total flowers is 10 to 9 + 10 = 19, or 10:19. Write the ratio as a fraction: $10:19 = \frac{10}{19}$. Set up a proportion. If 10 flowers of 19 total flowers are lilies, then 30 flowers of x total flowers are lilies: $\frac{10}{19} = \frac{30}{x}$. Cross multiply: $(10)(x) = (19)(30)$, $10x = 570$. Divide both sides of the equation by 10: $\frac{10x}{10} = \frac{570}{10}$, $x = 57$. If there are 30 lilies, then there are 57 total flowers.

23. If the ratio of vanilla pops to chocolate pops is 11:13, then the ratio of chocolate pops to total pops is 13 to 11 + 13 = 24, or 13:24. Write the ratio as a fraction: $13:24 = \frac{13}{24}$. Set up a proportion. If 13 pops of 24 total pops are chocolate, then x pops of 120 total pops are chocolate: $\frac{13}{24} = \frac{x}{120}$. Cross multiply: $(24)(x) = (13)(120)$, $24x = 1,560$. Divide both sides

of the equation by 24: $\frac{24x}{24} = \frac{1,560}{24}$, $x = 65$. If there are 120 total pops, then 65 of them are chocolate.

24. If the ratio of magazines to newspapers is 8:3, then the ratio of magazines to total magazines and newspapers is 8 to 8 + 3 = 11, or 8:11. Write the ratio as a fraction: $8:11 = \frac{8}{11}$. Set up a proportion. If 8 of 11 total magazines and newspapers are magazines, then 72 of x total magazines and newspapers are magazines: $\frac{8}{11} = \frac{72}{x}$. Cross multiply: $(8)(x) = (11)(72)$, $8x = 792$. Divide both sides of the equation by 8: $\frac{8x}{8} = \frac{792}{8}$, $x = 99$. If there are 72 magazines, then there are 99 total magazines and newspapers.

What's a Decimal?

WHAT'S AROUND THE BEND

- ➥ What Does the Word *Decimal* Mean?
- ➥ Learning about Place Value
- ➥ Naming Decimals
- ➥ Real-Life Decimals
- ➥ Rounding Decimals
- ➥ Comparing and Ordering Decimals

DECIMALS: TEN SYMBOLS

You probably learned how to count in kindergarten, or maybe even earlier than that. Counting to ten is easy now: 1, 2, 3, 4, 5, 6, 7, 8, 9, 10. But did you ever think about what those numbers mean? We count using the **decimal** system. We have ten symbols that we use to write numbers. It's easy to overlook the first number: 0.

FUEL FOR THOUGHT

A DECIMAL IS a number that is written using one or more of ten symbols. The ten decimal symbols are 0, 1, 2, 3, 4, 5, 6, 7, 8, and 9. These ten symbols can be used to write any number in the decimal system, such as 17 or 323,119. The prefix *dec-* means ten, which is why the decimal system is also called a *base-ten* system. Each digit in a decimal number is equal to that digit multiplied by a power of ten.

INSIDE TRACK

THERE ARE MANY kinds of number systems. For instance, the binary system is a base two system in which only two symbols, 0 and 1, are used to represent all numbers. The hexadecimal system is a number system that uses 16 symbols (six of which are letters!) to represent all numbers.

Every digit in a number has a place and that place has a value. Let's look at the number 323. The first 3 in the number does not have the same value as the second 3 in the number, because it is in a different place. What can we use to find the value of each 3 in the number 323? The place value system.

FUEL FOR THOUGHT

THE PLACE VALUE system uses a number system, such as the decimal system, and the position (or place) of each digit in a number to determine what value each digit in the number is worth.

A decimal number can have many digits. Each digit goes in a place and that place has a name. A **decimal point** is used to separate the parts of a number that are one or greater from the parts of the number that are worth between one and zero. First, we'll look at the digits to the left of the decimal point.

Left Side = Whole Numbers

The digits to the left of the decimal point are used to make up whole numbers. The value zero can also be expressed to the left of the decimal point. The value zero is written simply "0." That zero is written in the **ones place**, which is the first place to the left of the decimal point. Why is the first place to the left of the decimal point called the ones place? Because the value of a digit in this place is equal to that digit multiplied by 1. Whole numbers, or numbers that have no digits to the right of the decimal point, are often written without a decimal point. For instance, the number 7 could be written as 7.0, but we don't need to put a decimal point if every digit to the right of the decimal point is a 0.

Example

The number "3" has a value of three.

That sounds a little confusing. You probably never thought about why the number "3" has a value of 3. The number "3" has one digit, the digit 3, in the ones place. A number in the ones place must be multiplied by 1 to find its value: $3 \times 1 = 3$. Because there are no other digits in the number 3, the value of the number 3 is 3. This might STILL be confusing, but it becomes a little clearer when we look at larger numbers.

Example

The number "15" has a value of 15.

Again, you might be saying to yourself, "Of course 15 has a value of 15." But let's look at why. The number 15 has two digits: a 5 in the ones place, and a 1 in the tens place. The **tens place** is the second digit to the left of the decimal point. The value of a digit in this place is equal to that digit multiplied by 10. The digit 1 in the number 15 is equal to $1 \times 10 = 10$. The digit 5 in the number 15 is equal to $5 \times 1 = 5$. Now, we add the values of the two digits: $10 + 5 = 15$. And that is why 15 has a value of 15.

Example

The number "677" has a value of 677.

The number 677 has three digits: a 6 in the hundreds place, a 7 in the tens place, and a 7 in the ones place. The **hundreds place** is the third digit to the left of the decimal point. The value of a digit in this place is equal to that digit multiplied by 100. The digit 6 in the number 677 is equal to $6 \times 100 = 600$. The next digit, 7, is in the tens place. $7 \times 10 = 70$. The next digit is also a 7, but this 7 is in the ones place. It has a different value because it is in a different place. A 7 in the ones place has a value of $7 \times 1 = 7$. Now we add the value of each digit:

$$
\begin{array}{r}
600 \\
70 \\
+\ 7 \\
\hline
677
\end{array}
$$

The number "677" has a value of 677 because it has a 6 in the hundreds place, a 7 in the tens place, and a 7 in the ones place.

PRACTICE LAP

DIRECTIONS: Answer each question. You can check your answers at the end of this chapter.

  1. What is the value of the digit 8 in 428?
  2. What is the value of the digit 2 in 428?
  3. What is the value of the digit 4 in 428?
  4. In the number 546, in what place is the digit 5?
  5. In the number 721, in what place is the digit 2?
  6. In the number 843, what digit is in the tens place?
  7. In the number 987, what digit is in the ones place?

The following chart shows the values of each digit to the left of the decimal point all the way up to the billions place.

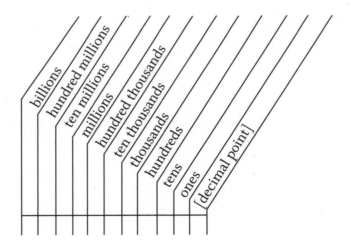

This next chart shows how to find the value of each digit in a decimal number:

Ones place	Multiply by 1
Tens place	Multiply by 10
Hundreds place	Multiply by 100
Thousands place	Multiply by 1,000
Ten thousands place	Multiply by 10,000
Hundred thousands place	Multiply by 100,000
Millions place	Multiply by 1,000,000
Ten millions place	Multiply by 10,000,000
Hundred millions place	Multiply by 100,000,000
Billions place	Multiply by 1,000,000,000

Every time we move one place farther to the left of the decimal point, the value of that place is ten times bigger. This is because the decimal system is a base-10 system. Look again at the preceding chart. The digit in the ones place is multiplied by 1, or 10^0. The digit in the tens place is multiplied by 10, or 10^1. The digit in the hundreds place is multiplied by 100, or 10^2, and so on. If a number were so big that it had a digit 17 places to the left of the decimal point, that value of that digit would be equal to that digit multiplied by 10^{17}!

Let's look at some bigger numbers.

Example

05,867,167

Put the number into the chart:

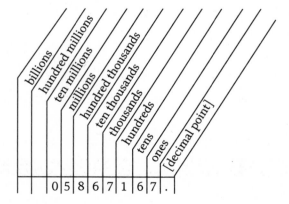

This number has a 0 in the ten millions place, a 5 in the millions place, an 8 in the hundred thousands place, a 6 in the ten thousands place, a 7 in the thousands place, a 1 in the hundreds place, a 6 in the tens place, and a 7 in the ones place. The values of each digit are:

$$0 \times 10,000,000 = 0$$
$$5 \times 1,000,000 = 5,000,000$$
$$8 \times 100,000 = 800,000$$
$$6 \times 10,000 = 60,000$$
$$7 \times 1,000 = 7,000$$
$$1 \times 100 = 100$$
$$6 \times 10 = 60$$
$$7 \times 1 = 7$$

The sum of these values is 5,867,167.

INSIDE TRACK

WHEN A ZERO appears to the left of the decimal point in front of all non-zero digits, it is called a "leading zero." For example, the number 054 has a leading zero. When a number has a leading zero, it can be ignored and left off the number. 054 = 54.

PRACTICE LAP

DIRECTIONS: Answer each question. You can check your answers at the end of this chapter.

8. What is the value of the digit 9 in 294,765,321?
9. What is the value of the digit 0 in 7,331,032?
10. In the number 32,487,906, in what place is the digit 4?
11. In the number 5,652,777,319, in what place is the digit 2?
12. In the number 765,382, what digit is in the hundred thousands place?
13. In the number 21,549,610, what digit is in the ten thousands place?

CAUTION!

WHILE LEADING ZEROS can be ignored, zeros within a number cannot be ignored. They hold an important place. Look at the numbers 406 and 46. The first number has a 4 in the hundreds place, a 0 in the tens place, and a 6 in the ones place. That number has a value much larger than 46, which has a 4 in the tens place and a 6 in the ones place. The middle 0 in 406 shows that although there are no tens, there are hundreds. If a number had 4 thousands, 0 hundreds, 0 tens, and 6 ones, it would be written as 4,006. Each middle zero holds a place and cannot be ignored.

Right Side = Part of a Whole Number

The digits to the right of the decimal point are part of a whole number. The first digit to the right of the decimal point is the **tenths place**. For instance, the number 0.7 has a value of seven tenths. The digit in the ones place is 0, and the digit in the tenths place is 7. How did we find the value of 0.7? The same way we found the value of whole numbers: by multiplying each digit by the value of its place. The value of a digit in the tenths place is equal to that digit multiplied by 0.1, or 10^{-1}; $7 \times 0.1 = 0.7$. The number 0.7 can also be written as just .7, because the 0 is a leading zero.

Example

The number "0.9" has a value of nine tenths.

The digit in the tenths place is multiplied by one tenth, 0.1: $9 \times 0.1 = 0.9$.

Example

The number "4.5" has a value of four and five tenths.

The digit in the ones place is multiplied by 1 and the digit in the tenths place is multiplied by 0.1: $4 \times 1 = 4$ and $5 \times 0.1 = 0.5$, which is why the number 4.5 has a value of four and five tenths.

The second digit to the right of the decimal point is the **hundredths place**. To find the value of a digit in the hundredths place, multiply it by one hundredth, or 0.01. One hundredth is the same as 10^{-2}. Numbers with negative exponents do not necessarily have a negative value, but they do have a value that is less than one.

FUEL FOR THOUGHT

THE TENTHS PLACE is the first place to the right of the decimal point. The **hundredths place** is the second place to the right of the decimal point, and the **thousandths place** is the third place to the right of the decimal point. As we move farther to the right of the decimal point, the value of each digit is ten times smaller.

The following chart now shows the values of each digit not just to the left of the decimal point, but also to the right of the decimal point, all the way down to the billionths place.

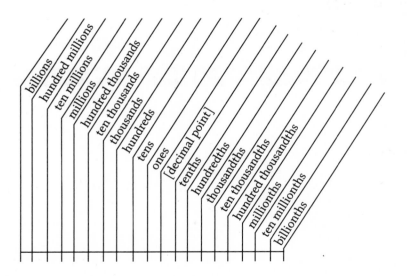

This next chart shows how to find the value of each digit to the right of the decimal point:

Tenth place	Multiply by 0.1, 10^{-1}
Hundredths place	Multiply by 0.01, or 10^{-2}
Thousandths place	Multiply by 0.001, or 10^{-3}
Ten thousandths place	Multiply by 0.0001, or 10^{-4}
Hundred thousandths place	Multiply by 0.00001 or 10^{-5}
Millionths place	Multiply by 0.000001, or 10^{-6}
Ten millionths place	Multiply by 0.0000001, or 10^{-7}
Hundred millionths place	Multiply by 0.00000001, or 10^{-8}
Billionths place	Multiply by 0.000000001, or 10^{-9}

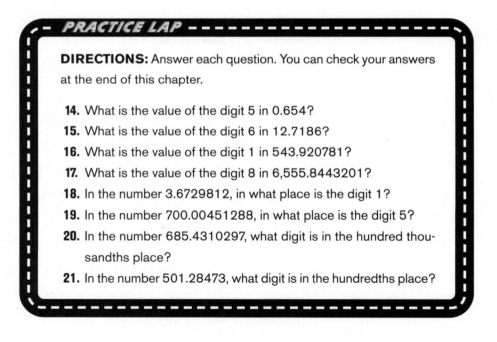

PRACTICE LAP

DIRECTIONS: Answer each question. You can check your answers at the end of this chapter.

14. What is the value of the digit 5 in 0.654?

15. What is the value of the digit 6 in 12.7186?

16. What is the value of the digit 1 in 543.920781?

17. What is the value of the digit 8 in 6,555.8443201?

18. In the number 3.6729812, in what place is the digit 1?

19. In the number 700.00451288, in what place is the digit 5?

20. In the number 685.4310297, what digit is in the hundred thousandths place?

21. In the number 501.28473, what digit is in the hundredths place?

CAUTION!

LOOK CLOSELY AT the place value chart. You'll see that there is a tens place on the left side of the decimal point and a tenths place on the right side of the decimal point. You'll also see a hundreds place on the left side and a hundredths place on the right side, and so on. However, these pairs are not the same distance (number of places) from the decimal point. The second digit to the left of the decimal point is the tens place, but the *first* digit to the right of the decimal point is the tenths place. The third digit to the left of the decimal point is the hundreds place, but the second digit to the right of the decimal point is the hundredths place. This is because there is no "oneths" place. So, be careful when reading your decimals. The digit six places to the left of the decimal point does not have a name similar to the digit six places to the right of the decimal point.

Now that we understand the value of each place in a decimal number, let's learn how to name decimals. We will start with naming whole numbers, because you're probably already pretty familiar with those!

Example

453

This number has a 4 in the hundreds place, a 5 in the tens place, and a 3 in the ones place. We read this decimal number as "four hundred fifty-three." The digits are read from left to right. Digits in the tens and ones place are hyphenated for numbers greater than twenty that have a non-zero ones digit.

Example

5,988,127

This number has a 5 in the millions place, a 9 in the hundred thousands place, an 8 in the ten thousands place, an 8 in the thousands place, a 1 in the hundreds place, a 2 in the tens place, and a 7 in the ones place. It is read as "Five million, nine hundred eighty-eight thousand, one hundred twenty-seven." Just as we put commas after every three digits when we write a number out in digits, we name three digits at a time, followed by a comma.

Example

267.544

This number has a 2 in the hundreds place, a 6 in the tens place, a 7 in the ones place, a 5 in the tenths place, a 4 in the hundredths place, and a 4 in the thousandths place. It is read as "Two hundred sixty-seven and five hundred forty-four thousandths." The decimal point is read as *and*. We read three digits at a time on each side of the decimal point, referring to .544 as "five hundred forty-four thousandths" rather than "five tenths, four hundredths, and four thousandths."

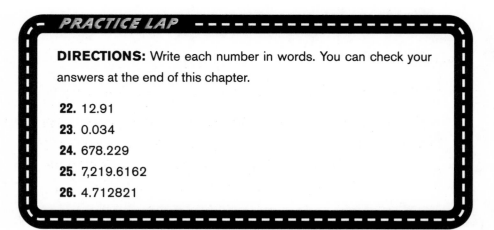

PRACTICE LAP

DIRECTIONS: Write each number in words. You can check your answers at the end of this chapter.

22. 12.91

23. 0.034

24. 678.229

25. 7,219.6162

26. 4.712821

CAUTION!

LEADING ZEROS CAN be tricky. The zero in 034.78 is a leading zero and can be ignored. However, the number 0.034 has only ONE leading zero. 0.034 is the same as .034, but it is NOT the same as 0.34. Zeros between the decimal point and other non-zero digits to the right of the decimal point are **placeholders**. The zero in 5.021 is as important as the zero in 5,021. It holds a place in the number, and cannot be ignored. However, trailing zeros to the right of the decimal point can be ignored. The number 6.670 is the same as the number 6.67. In fact, 6.670000000 is the same as 6.67 too, which is also the same as 00000006.67. Leading zeros to the left of the decimal point and trailing zeros to the right of the decimal point can be ignored—but all other zeros cannot be ignored.

Example

Write the number "eighty-one and nine tenths" in digits.

Eighty-one is 81. The word *and* means that the decimal point comes next, followed by nine tenths, or 0.9. Eighty-one and nine tenths is 81.9.

Example

Write the number "three hundred and eight thousandths" in digits.

Three hundred is 300, because there are no tens or ones in this number. There are also no tenths or hundredths, so we will put zeros in those places as well. Eight thousandths is 0.008, so the number three hundred and eight thousandths is 300.008.

DIRECTIONS: Write each number in digits. You can check your answers at the end of this chapter.

27. thirty-three and five tenths
28. nine and two hundred forty-two ten-thousandths
29. seven thousand, eight hundred forty-one and sixteen hundredths
30. eleven and three millionths

USING DECIMALS

Now that we know what decimals are, how can we use them? Think about a meter stick. A meter is about 3.28 feet, which is a little longer than a yard. A typical door is about 2.5 meters tall. That means that two full meter sticks plus five tenths of a meter stick, placed end to end, would be the same height as the door. Without decimals, we would have to say that the door was two meters tall or three meters tall. Decimals allow us to be more precise with our measurements.

Example

Estimate the length of a baseball bat.

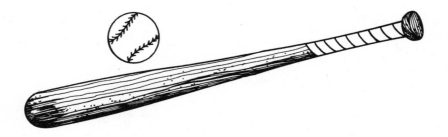

A baseball bat is almost one meter long. A good estimate for the length of a bat might be 0.8 meters, or maybe 0.76 meters.

CAUTION!

DECIMALS IN REAL-LIFE situations must have a units label in order to have meaning. The number 0.76 without a label could be any length: It could be 0.76 meters, the length of a bat, or it could be 0.76 centimeters—half the length of a dime! If you are using decimals for measurements, always include a label.

PACE YOURSELF

TAKE A METER stick and measure five objects in meters. Write your answers to two decimal places. For instance, you might find that your television remote control is 0.18 meters long and your bed is 1.72 meters long. Save these measurements—we'll use them in a little while!

CLOSE ENOUGH: ROUNDING

We've looked at decimals that were so precise that they had digits to the millionths place. But sometimes, we don't need a decimal to be THAT exact. In these situations, we round to a certain place.

FUEL FOR THOUGHT

ROUNDING IS THE process of taking a number and making it less precise by removing one or more digits from the end of the number, replacing those digits with zeros if necessary. The digit in the place to which you are rounding either is increased by one or stays the same according to the rules of rounding.

We can round a number to any place. To round a number to a certain place value, look at the digit to the immediate right. If that digit is 5 or greater, we round up. If that digit is less than 5, we round down.

Example

Round 463 to the tens place.

To round the number 463 to the tens place, we look at the digit to the immediate right of the tens place: the digit in the ones place. Because the digit in the ones place, 3, is less than 5, we round down. Because we are rounding to the tens place, and we know that we are rounding down, the tens digit stays the same and all digits to the right of the tens place become zero. In the number 463, only one digit is to the right of the tens place. The 3 in 463 becomes 0. The number 463, when rounded to the tens place, becomes 460.

What if we wanted to round 463 to the hundreds place? We would look at the digit to the immediate right of the hundreds place: the tens place. Because the digit in the tens place, 6, is greater than 5, we round up. Because we are rounding to the hundreds place, and we know that we are rounding up, the hundreds digit increases by one and all digits to the right of the hundreds place become zero. In the number 463, there are two digits to the right of the hundreds place. After changing the 4 in the hundreds place to 5, the 6 and the 3 in 463 become 0. The number 463, when rounded to the hundreds place, becomes 500.

Now let's look at rounding a number with a few more decimal places.

Example

Round 7.28 to the nearest tenth.

To round the number 7.28 to the tenths place, we look at the digit to the immediate right of the tenths place: the digit in the hundredths place. Because the digit in the hundredths place, 8, is greater than 5, we round up. The tenths digit increases by one, from 2 to 3, and the hundredths digit becomes 0. 7.28 rounded to the nearest tenth is 7.30, or 7.3.

What if we rounded 7.28 to the nearest ones place? We'd look at the digit in the tenths place, because that is the place to the immediate right of the ones place. There is a 2 in the tenths place, so we must round down. The digit in the ones place, 7, stays the same and all digits to the right become zero. 7.28 rounded to the nearest one is 7.00. Because those zeros are trailing zeros, we can say that 7.28 rounded to the nearest one is 7.

Example

Round 574.299199 to the nearest thousandth.

The digit to the immediate right of the thousandths place is the digit in the ten thousandths place, 1. Because 1 is less than 5, we round down. Notice that even though the thousandths digit itself is greater than 5, and even though the digits to the right of the ten thousandths place are greater than 5, we still round down, because we are rounding to the thousandths place. The thousandths digit, 9, stays the same and the digits to the right become zero. 574.299199 to the nearest thousandth is 574.299000, or 574.299.

CAUTION!

BE SURE NEVER to "chain round," or use a rounded version of a number to round again. Here's why:

Look at the number 8.47. What is 8.47 rounded to the nearest tenth? Because the hundredths digit, 7, is greater than 5, we increase the tenths digit by one and change the digits to the right of the tenths place to zero. 8.47 rounded to the nearest tenth is 8.5. Now, what is 8.47 rounded to the nearest one? Even though you know that 8.47 rounded to the nearest tenth is 8.5, do not use the number 8.5. The number 8.5 has a 5 in the tenths place. If we were to round the number 8.5 to the nearest one, we would round up, because 5 is equal to 5. We would increase the ones digit from 8 to 9, and change the digits to the right to zero. 8.5 to the nearest one is 9.0, or 9. However, 8.47 rounded to the nearest one is not 9. The tenths digit of 8.47 is 4, which means that we must round down. 8.47 to the nearest one is 8. That is why you must never round a number twice.

To recap:

8.47 rounded to the nearest tenth is 8.5.
8.5 rounded to the nearest one is 9.
8.47 rounded to the nearest one is 8.

Sometimes when we round, we have to increase more than just the digit in the place in which we're rounding. This happens when we need to round up, and the digit to be rounded up is a 9.

Example

Round 5.896 to the nearest hundredth.

We begin by looking at the hundredths digit, 9, and the digit to the immediate right of the hundredths digit, the thousandths digit, which is 6. Because 6 is greater than 5, we increase the hundredths digit by 1 and change the digits to the right to zero. However, when we increase 9 by 1, it becomes 10. When this happens, we must increase by 1 the digit to the LEFT of the place in which we're rounding. In the number 5.896, there is an 8 in the tenths place. This 8 becomes a 9, and the digits to the right of it become 0. 5.896 rounded to the nearest hundredth is 5.900, or 5.9. When 5.896 is rounded to the nearest hundredth, the hundredths digit disappears entirely! In fact, the number 5.896, when rounded to the nearest hundredth, is exactly the same as when it is rounded to the nearest tenth—try it out yourself!

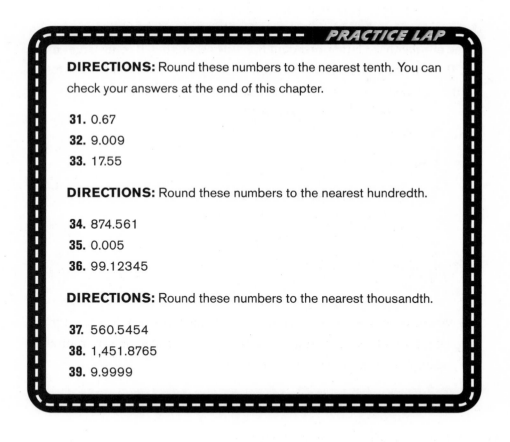

PRACTICE LAP

DIRECTIONS: Round these numbers to the nearest tenth. You can check your answers at the end of this chapter.

31. 0.67
32. 9.009
33. 17.55

DIRECTIONS: Round these numbers to the nearest hundredth.

34. 874.561
35. 0.005
36. 99.12345

DIRECTIONS: Round these numbers to the nearest thousandth.

37. 560.5454
38. 1,451.8765
39. 9.9999

PACE YOURSELF

FIND THE FIVE measurements you wrote down earlier in this chapter. Now, round each measurement to the hundredths place, the tenths place, and the ones place. Be sure to round your original measurement each time—don't round a rounded number!

WHICH IS BIGGER?

As we move left from one digit to the next in a decimal number, the place value increases by a power of 10. We can compare two decimals by lining up their decimal points and comparing corresponding digits from left to right. If one number has a larger first digit than the other, then it is the larger number. If the two numbers have the same first digit, we move one place to the right and compare again. Remember, line up the decimal points first.

Example

43.98 versus 34.98

First, line up the decimal points:

43.98
34.98

Next, compare the digits from left to right. We begin in the tens place. The first number, 43.98, has a 4 in the tens place. The second number, 34.98, has a 3 in the tens place. Because 4 is greater than 3, 43.98 is greater than 34.98—we don't need to look at any other digits.

Example

677.55 versus 687.01

First, line up the decimal points:

677.55
687.01

Next, compare the digits from left to right. We begin in the hundreds place. The first number, 677.55, has a 6 in the hundreds place. The second number, 687.01, has a 6 in the hundreds place too. Because both numbers have the same digit in the hundreds place, we move one place to the right and compare again. The first number has a 7 in the tens place and the second number has an 8 in the tens place. Because 8 is greater than 7, 687.01 is greater than 677.55. Notice that it doesn't matter that the decimal part of 677.55 is larger than the decimal part of 687.01. We compare digits from left to right, and as soon as we find different digits between the two numbers, we're ready to make our decision.

Example

74.67 versus 100.31

First, line up the decimal points:

74.67
100.31

Notice that the first number, 74.67, does not have a digit in the hundreds place. Because 100.31 has a digit in the hundreds place that isn't zero, it must be the larger number.

INSIDE TRACK

WHEN COMPARING TWO numbers that have a different number of digits, use zeros to fill in each number so that it is easier to compare them. For instance, compare 5.607 to 44.1. Lining up the decimal points gives us:

    5.607
    44.1

To make the comparison easier, add a leading zero to 5.607 and two trailing zeros to 44.1:

    05.607
    44.100

Now it's easier to see that the second number, 44.1, has a 4 in the tens place, and the first number, 5.607, has a zero in the tens place. 44.1 has fewer digits, but it is the bigger number.

Example

8.52376 versus 8.5276

First, line up the decimal points and add a trailing zero to the second number:

    8.52376
    8.52760

Next, compare the digits from left to right. We begin in the ones place. The first number, 8.52376, has an 8 in the ones place. The second number, 8.52760, also has an 8 in the ones place. We move one place to the right and compare again. The first number has a 5 in the tenths place and the second number has a 5 in the tenths place. Because the digits are the same, we move one place to the right again. Both numbers have a 2 in the hundredths place.

Again, we move one place to the right. The first number has a 3 in the thousandths place and the second number has a 7 in the thousandths place. Because 7 is greater than 3, 8.5276 is greater than 8.52376.

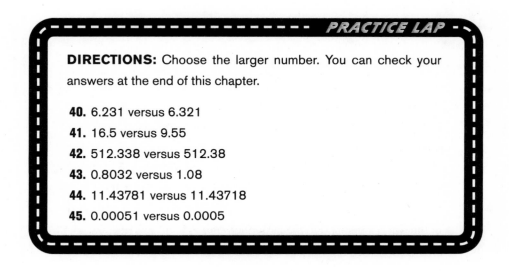

PRACTICE LAP

DIRECTIONS: Choose the larger number. You can check your answers at the end of this chapter.

40. 6.231 versus 6.321

41. 16.5 versus 9.55

42. 512.338 versus 512.38

43. 0.8032 versus 1.08

44. 11.43781 versus 11.43718

45. 0.00051 versus 0.0005

Ordering a Set of Decimals

Now that we know how to compare two decimals, we can compare a set of decimals and put that set in order from greatest to least. How? The same way we found the larger of two decimals: by lining up the decimal points of each number, adding leading and trailing zeros, and then comparing digits from left to right.

Example

4.712, 4.172, 4.271

Start by lining up the decimal points of the three numbers:

4.712
4.172
4.271

Next, compare the digits from left to right. Each number has a 4 in the ones place, so we move to the tenths place. The first number has a 7 in the tenths place, the second number has a 1 in the tenths place, and the third number has a 2 in the tenths place. Because 7 > 2 > 1, the first number is the largest, followed by the third number, and then the second number.

FUEL FOR THOUGHT

WHEN COMPARING NUMBERS, there are three symbols that we use: the **greater than** sign (>), the **less than** sign (<), and the **equals sign** (=). If two numbers are the same, we use the equals sign. If the two numbers are different, we use either the greater than sign or the less than sign. The open end of the symbol always goes with the larger number. Because 8 is greater than 6, we write 8 > 6, with the open end of the symbol pointing toward the larger number, 8. We could also write 6 < 8, because this comparison also shows the open end of the symbol pointing to the larger number.

Example

53.2495, 53.248, 53.312, 53.2488, 53.32

Start by lining up the decimal points of the three numbers and adding trailing zeros as needed:

53.2495
53.2480
53.3120
53.2488
53.3200

Compare the digits from left to right. Each number has a 5 in the tens place and a 3 in the ones place, so we move to the tenths place. The first two numbers have a 2 in the tenths place, as does the fourth number. The third number and the fifth number have a 3 in the tenths place. So, the third number and the fifth number are both larger than the first, second, and fourth numbers.

Now look just at the third and fifth numbers. The third number, 53.3120, has a 1 in the hundredths place, and the fifth number, 53.3200, has a 2 in the hundredths place. Because $2 > 1$, the fifth number is larger than the third number. We know now that the fifth number is the largest number, and the third number is the next largest number.

Look at the first, second, and fourth numbers again. They all have the same tenths digit, and they all have the same hundredths digit. Move to the thousandths place. The first number, 53.2495, has a 9 in the thousandths place, while the second number, 53.2480 and the fourth number, 53.2488, both have an 8 in the thousandths place. Because $9 > 8$, the first number is larger than the second and fourth numbers.

So far, our order from largest to smallest is 53.3200, 53.3120, 53.2495. Now let's look at just the second and fourth numbers. They both have the same thousandths digit. Move to the ten thousandths place. The second number, 53.2480, has a 0 in the ten thousandths place, while the fourth number, 53.2488, has an 8 in the ten thousandths place. Because $8 > 0$, the fourth number is larger than the second number.

Finally, we have our set in order from greatest to least: $53.3200 > 53.3120 > 53.2495 > 53.2488 > 53.2480$. We can also write the set in order from least to greatest: $53.2480 < 53.2488 < 53.2495$ $53.3120 < 53.3200$.

CAUTION!

WHEN ORDERING NUMBERS, always read the directions carefully. Sometimes you will be asked to order numbers from least to greatest, and sometimes you will be asked to order numbers from greatest to least. Be sure to order your numbers in the manner in which the directions state.

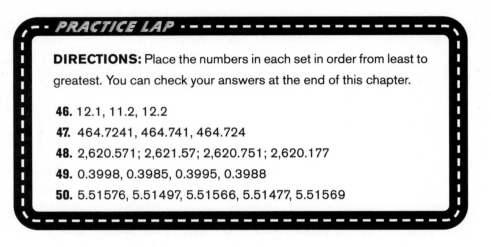

- - • *PRACTICE LAP* • - - - - - - - - - - - - - - - - - - -

DIRECTIONS: Place the numbers in each set in order from least to greatest. You can check your answers at the end of this chapter.

46. 12.1, 11.2, 12.2

47. 464.7241, 464.741, 464.724

48. 2,620.571; 2,621.57; 2,620.751; 2,620.177

49. 0.3998, 0.3985, 0.3995, 0.3988

50. 5.51576, 5.51497, 5.51566, 5.51477, 5.51569

We now understand what decimal numbers are and how they are formed. Rounding is a skill we will use to shorten our answers to upcoming decimal questions. Next up: basic operations with decimals.

ANSWERS

1. The 8 is in the ones place, so it has a value of $8 \times 1 = 8$.

2. The 2 is in the tens place, so it has a value of $2 \times 10 = 20$.

3. The 4 is in the hundreds place, so it has a value of $4 \times 100 = 400$.

4. The digit 5 in the number 546 is in the third place to the left of the decimal point, which is the hundreds place.

5. The digit 2 in the number 721 is in the second place to the left of the decimal point, which is the tens place.

6. The tens place is the second place to the left of the decimal point. In the number 843, the 4 is the digit in the second place to the left of the decimal point.

7. The ones place is the first place to the left of the decimal point. In the number 987, the 7 is the digit in the first place to the left of the decimal point.

8. The 9 is in the ten millions place, so it has a value of $9 \times 10,000,000 = 90,000,000$.

9. The 0 is in the hundreds place, so it has a value of $0 \times 100 = 0$.

10. The digit 4 in the number 32,487,906 is in the sixth place to the left of the decimal point, which is the hundred thousands place.

11. The digit 2 in the number 5,652,777,319 is in the seventh place to the left of the decimal point, which is the millions place.

12. The hundred thousands place is the sixth place to the left of the decimal point. In the number 765,382, the 7 is the digit in the sixth place to the left of the decimal point.

13. The ten thousands place is the fifth place to the left of the decimal point. In the number 21,549,610, the 4 is the digit in the fifth place to the left of the decimal point.

14. The 5 is in the hundredths place, so it has a value of $5 \times 0.01 = 0.05$.

15. The 6 is in the ten thousandths place, so it has a value of $6 \times 0.0001 = 0.0006$.

16. The 1 is in the millionths place, so it has a value of $1 \times 0.000001 = 0.000001$.

17. The 8 is in the tenths place, so it has a value of $8 \times 0.1 = 0.8$.

18. The digit 1 in the number 3.6729812 is in the sixth place to the right of the decimal point, which is the millionths place.

19. The digit 5 in the number 700.00451288 is in the fourth place to the right of the decimal point, which is the ten thousandths place.

20. The hundred thousandths place is the fifth place to the right of the decimal point. In the number 685.4310297, the 2 is the digit in the fifth place to the right of the decimal point.

21. The hundredths place is the second place to the right of the decimal point. In the number 501.28473, the 8 is the digit in the second place to the right of the decimal point.

22. twelve and ninety-one hundredths

23. thirty-four thousandths

24. six hundred seventy-eight and two hundred twenty-nine thousandths

25. seven thousand, two hundred nineteen and six thousand, one hundred sixty-two thousandths

26. four and seven hundred twelve thousand, eight hundred twenty-one millionths

27. 33.5

28. 9.0242

29. 7,841.16

30. 11.000003

31. The digit to the immediate right of the tenths place is the digit in the hundredths place, 7. Because 7 is greater than 5, we round up. The tenths digit increases by 1 to 7, and the digits to the right become zero. 0.67 to the nearest tenth is 0.70, or 0.7.

32. The digit to the immediate right of the tenths place is the digit in the hundredths place, 0. Because 0 is less than 5, we round down. The tenths digit stays at 0, and the digits to the right become zero. 9.009 to the nearest tenth is 9.000, or 9.

33. The digit to the immediate right of the tenths place is the digit in the hundredths place, 5. Because 5 is equal to 5, we round up. The tenths digit increases by 1 to 6, and the digits to the right become zero. 17.55 to the nearest tenth is 17.60, or 17.6.

34. The digit to the immediate right of the hundredths place is the digit in the thousandths place, 1. Because 1 is less than 5, we round down. The hundredths digit stays at 6, and the digits to the right become zero. 874.561 to the nearest hundredth is 874.560, or 874.56.

35. The digit to the immediate right of the hundredths place is the digit in the thousandths place, 5. Because 5 is equal to 5, we round up. The hundredths digit increases by 1 to 1, and the digits to the right become zero. 0.005 to the nearest hundredth is 0.010, or 0.01.

36. The digit to the immediate right of the hundredths place is the digit in the thousandths place, 3. Because 3 is less than 5, we round down. The hundredths digit stays at 2, and the digits to the right become zero. 99.12345 to the nearest hundredth is 99.12000, or 99.12.

37. The digit to the immediate right of the thousandths place is the digit in the ten thousandths place, 4. Because 4 is less than 5, we round down. The thousandths digit stays at 5, and the digits to the right become zero. 560.5454 to the nearest thousandth is 560.5450, or 560.545.

38. The digit to the immediate right of the thousandths place is the digit in the ten thousandths place, 5. Because 5 is equal to 5, we round up. The thousandths digit increases by 1 to 7, and the digits to the right become zero. 1,451.8765 to the nearest thousandth is 1,451.8770, or 1,451.877.

39. The digit to the immediate right of the thousandths place is the digit in the ten thousandths place, 9. Because 9 is greater than 5, we round up. The thousandths digit increases by 1 to 10, which means that we must increase the digit to the left, the tenths digit. That digit increases from 9 to 10, which means that we must increase the digit to its left, the ones digit. The ones digit increases from 9 to 10, which means that the tens digit increase by one. The digits to the right of the tens digit become zero; 9.9999 to the nearest thousandth is 10.0000, or 10.

40. Both numbers have a 6 in the ones place, so we move right to the tenths digit. The first number has a 2 in the tenths place and the second number has a 3 in the tenths place, so the second number is bigger; 6.321 is greater than 6.231.

41. Add a trailing zero to the first number and a leading zero to the second number. The first number has a 1 in the tens place and the second number has a 0 in the tens place, so the first number is bigger; 16.5 is greater than 9.55.

42. Add a trailing zero to the second number. The numbers have the same digits in the hundreds, tens, ones, and tenths places. The first number has a 3 in the hundredths place and the second number has an 8 in the hundredths place, so the second number is bigger; 512.38 is greater than 512.338.

43. Add two trailing zeros to the second number. The first number has a 0 in the ones place and the second number has a 1 in the ones place, so the second number is bigger; 1.08 is greater than 0.8032.

44. The numbers have the same digits in the tens, ones, tenths, hundredths, and thousandths places. The first number has an 8 in the ten thousandths place and the second number has a 1 in the ten thousandths place, so the first number is bigger; 11.43781 is greater than 11.43718.

45. Add a trailing zero to the second number. The numbers have the same digits in the ones, tenths, hundredths, thousandths, and ten thousandths places. The first number has a 1 in the hundred thousandths place and the second number has a 0 in the hundred thousandths place, so the first number is bigger; 0.00051 is greater than 0.0005.

46. All three numbers have the same tens digit, so move to the ones place. The first and the third number have a 2 in the ones place, and the second number has a 1 in the ones place. Because $2 > 1$, the second number is the least number. Look again at the first and third numbers. They have the same ones digit, so compare the tenths digits. The first number has a 1 in the tenths place and the third number has a 2 in the tenths place. Because $2 > 1$, the first number is less than the third number. The set, in order from least to greatest, is 11.2, 12.1, 12.2.

47. All three numbers have the same hundreds, tens, ones, and tenths digits. Compare the hundredths digits of each number. The first and third numbers have a 2 in the hundredths place and the second number has a 4 in the hundredths place. Because $4 > 2$, the second number is the greatest number. Now compare the thousandths digits of the first and third numbers. Because both numbers have a 4 in the thousandths place, move to the ten thousandths place. The first number has a 1 in the ten thousandths place and the third number has a 0 in the ten thousandths place. Because $1 > 0$, the first number is larger than the third number. The set, in order from least to greatest, is 464.724, 464.7241, 464.741.

48. All four numbers have the same thousands, hundreds, and tens digits. Compare the ones digits of each number. The first, third, and fourth numbers have a 0 in the ones place and the second number has a 1 in the ones place. Because $1 > 0$, the second number is the greatest number. Now compare the tenths digits of the first, third, and fourth numbers. The first number has a 5 in the tenths place, the third number has a 7 in the tenths place, and the fourth number has a 1 in the tenths place. Because $7 > 5 > 1$, the third number is greater than the first number, and the first number is greater than the fourth number. The set, in order from least to greatest, is 2,620.177; 2,620.571; 2,620.751; 2,621.57.

49. All four numbers have the same ones, tenths, and hundredths digits. Compare the thousandths digits of each number. The first and third numbers have a 9 in the thousandths place and the second and fourth numbers have an 8 in the thousandths place. Because $9 > 8$, the first and third numbers are larger than the second and fourth numbers. Look at the ten thousandths place of the second and fourth numbers. The ten thousandths digit of the second number is 5 and the ten thousandths digit of the fourth number is 8. Because 8 > 5, the second number is the lowest number, followed by the fourth number. Look again at the first and third numbers. The ten thousandths digit of the first number is 8 and the ten thousandths digit of the third number is 5. Because 8 > 5, the first number is the greatest number, followed by the third number. The set, in order from least to greatest, is 0.3985, 0.3988, 0.3995, 0.3998.

50. All five numbers have the same ones, tenths, and hundredths digits. Compare the thousandths digits of each number. The first, third, and fifth numbers have a 5 in the thousandths place and the second and fourth numbers have a 4 in the thousandths place. Because 5 > 4, the first, third, and fifth numbers are larger than the second and fourth numbers. Look at the ten thousandths place of the second and fourth numbers. The ten thousandths digit of the second number is 9 and the ten thousandths digit of the fourth number is 7. Because 9 > 7, the fourth number is the lowest number, followed by the second number. Look again at the first, third, and fifth numbers. The ten thousandths digit of the first number is 7 and the ten-thousandths digit of the third and fifth numbers is 6. Because 7 > 6, the first number is the greatest number. Now compare the hundred thousandths digits of the third and fifth numbers. The hundred thousandths digit of the third number is 6 and the hundred thousandths digit of the fifth number is 9. Because 9 > 6, the third number is less than the fifth number. The set, in order from least to greatest, is 5.51477, 5.51497, 5.51566, 5.51569, 5.51576.

8 Decimal Basics

In this chapter, we'll look at the four basic operations (addition, subtraction, multiplication, and division) with decimals and whole numbers. Remember, a **whole number** is a positive number that has no fractional part—no digits to the right of the decimal point.

ADDING DECIMALS TO WHOLE NUMBERS

Adding decimals to whole numbers, or even adding decimals to decimals, is a lot easier than adding fractions or mixed numbers. No common denominators, no simplifying answers. Just line up your columns and add!

Example

5 + 6.34

First, write the addition problem vertically, and put a decimal point on the end of any whole numbers. Then, place trailing zeros next to that decimal point until the whole number has the same number of digits as the decimal number:

$$
\begin{array}{r}
5.00 \\
+\ 6.34 \\
\hline
\end{array}
$$

Now, add column by column, just as you would add the numbers 500 and 634, and carry the decimal point down into your answer.

$$
\begin{array}{r}
5.00 \\
+\ 6.34 \\
\hline
11.34
\end{array}
$$

The answer to an addition problem involving one or more decimals will have as many digits to the right of the decimal point as the **addend** with the most digits to the right of the decimal point. In other words, if one number being added has three digits to the right of the decimal point, and the other has two digits to the right, then the answer will have three digits to the right of the decimal point. In the example above, one addend had zero places to the right of the decimal point, and the other addend had two places to the right of the decimal point. So, our answer had two places to the right of the decimal point. The only exception to this rule is if the last digit (rightmost digit) is a 0 after adding. Trailing zeros, as you know, can be removed.

FUEL FOR THOUGHT

AN ADDEND IS a number that is being added to another number. In the problem 5 + 6.34, both 5 and 6.34 are addends.

Example

918 + 45.8664

Write the addition problem vertically. Put a decimal point on the end of the whole number and place trailing zeros next to it:

918.0000
+ 45.8664

Add column by column and carry the decimal point down into your answer.

918.0000
+ 45.8664
963.8664

INSIDE TRACK

IF A WHOLE number is added to a decimal number that is between 0 and 1, just join the two numbers at the decimal point. The sum of 154 and 0.5874 is 154.5874—no addition required!

PRACTICE LAP

DIRECTIONS: Add the numbers below. You can check your answers at the end of this chapter.

1. 54 + 2.35
2. 27.568 + 3
3. 289 + 0.051
4. 74.65722 + 1,045
5. 874 + 8.74

ADDING DECIMALS TO DECIMALS

Adding decimals to decimals is just like adding decimals to whole numbers. It is just as important to line up your columns and add trailing zeros as needed.

Example

10.25 + 25.61

Again, write the problem vertically, lining up the decimal points:

$$
\begin{array}{r}
10.25 \\
+\ 25.61 \\
\hline
\end{array}
$$

Add column by column and carry the decimal point down into your answer.

$$
\begin{array}{r}
10.25 \\
+\ 25.61 \\
\hline
35.86
\end{array}
$$

Sometimes the hardest part of an addition problem with decimals is keeping the decimal points straight. After you have written an addition problem vertically, you should able to draw a straight line down through the decimal points of the addends and the answer.

Example

145.785 + 2.3

The first addend has six digits: three to the left of the decimal point and three to the right. The second addend has only two digits, one to the left of the decimal point and one to the right. Keep your columns straight and place two trailing zeros on the end of 2.3:

$$
\begin{array}{r}
145.785 \\
+\ 2.300 \\
\hline
\end{array}
$$

Add column by column and carry the decimal point down into your answer.

$$145.785$$
$$+\ \ 2.300$$
$$148.085$$

PACE YOURSELF

FOR EACH DECIMAL addition (or subtraction) problem that you do, take a pencil and draw a straight line through the decimal points of each number. This will help you check your answer. Drawing a straight line through the decimal points of the last answer shows us that we've kept our columns lined up correctly:

$$145\,|\,785$$
$$+\ \ \ 2\,|\,300$$
$$148\,|\,085$$

It is especially important to keep your columns straight when a decimal problem involves carrying.

Example

74.52 + 373.9

Line up the columns and add a trailing zero to 373.9. Then, add column by column:

$$74.52$$
$$+\,373.90$$
$$448.42$$

Don't forget to draw a line through the decimal points!

CAUTION!

THERE IS ONE case where the sum of two decimals has fewer dig-its to the right of the decimal point than the addend with the great-est number of digits to the right of the decimal point. That is when the sum of the digits in the rightmost column is equal to 10:

$$
\begin{array}{r}
5.64 \\
+\,12.96 \\
\hline
18.60
\end{array}
$$

The answer 18.60 can be written with one decimal place, as 18.6.

PRACTICE LAP

DIRECTIONS: Add the numbers below. You can check your answers at the end of this chapter.

6. 85.6 + 0.66
7. 108.47 + 8.4
8. 7.91 + 6,327.7
9. 19.159 + 207.44
10. 62.3906 + 58.906

SUBTRACTING DECIMALS FROM WHOLE NUMBERS

When we added decimals to whole numbers, we placed trailing zeros at the end of the whole numbers to help us keep our columns straight. When we sub-tract decimals from whole numbers, we *have* to add trailing zeros to the end of the whole numbers. We will also have to borrow, likely more than once.

Example

44 – 3.35

As with addition, we begin by lining up our columns. Place two trailing zeros on the end of the minuend (44):

$$44.00$$
$$\underline{- 3.35}$$

In order to do subtraction in the hundredths place, we must borrow from the tenths place. Because the tenths place also has a 0 in it, we must borrow from the ones place. Whenever we are subtracting a decimal from a whole number, we will have to borrow from at least the ones place, and rename every trailing zero we placed at the end of the whole number.

$$44.00$$
$$\underline{- 3.35}$$
$$40.65$$

As with addition, the answer (or difference) to our subtraction problem has the same number of digits to the right of the decimal point as the **operand** (minuend or subtrahend) with the greatest number of digits to the right of the decimal point. And as with addition, we should be able to draw a straight line through the decimal points of the minuend, subtrahend, and difference.

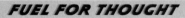

FUEL FOR THOUGHT

IN A SUBTRACTION problem, the **minuend** is the number from which you are subtracting. The **subtrahend** is the number you are subtracting, and the **difference** is the result of the subtraction. In the problem 4 − 3 = 1, 4 is the minuend, 3 is the subtrahend, and 1 is the difference. The numbers on which an operation (such as subtraction) is performed are called the **operands.** In the problem 4 − 3 = 1, 4 and 3 are operands.

Example

10 – 0.64381

Line up the decimal points and place five trailing zeros on the end of 10. Then, subtract:

$$
\begin{array}{r}
10.00000 \\
- 0.64381 \\
\hline
9.35619
\end{array}
$$

INSIDE TRACK

IN A PROBLEM like the previous one where a decimal with many digits to the right of the decimal point is being subtracted from a whole number, you can save time borrowing if you think about addition facts that add to 9. Look at each decimal place in the subtraction problem. The sum of the digit in the subtrahend and the digit in the difference is 9 in the tenths place, hundredths place, thousandths place, and ten thousandths place:

$$
\begin{array}{r}
10.00000 \\
- 0.64381 \\
\hline
9.35619
\end{array}
$$

The sum of the digits in the rightmost column, the hundred thousandths place, is 10. The digits to the right of a decimal point in the difference between a whole number and a decimal number always follow this pattern. If you are faster with addition facts, you can solve problems like these more easily, especially if the problem has A LOT of decimal places:

$$
\begin{array}{r}
14.000000000000000 \\
-1.679687548132048 \\
\hline
12.320312451867952
\end{array}
$$

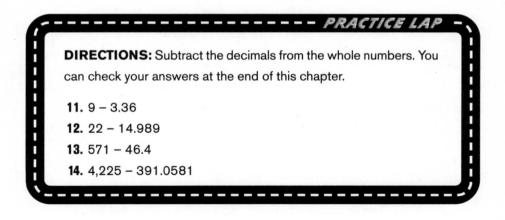

DIRECTIONS: Subtract the decimals from the whole numbers. You can check your answers at the end of this chapter.

11. 9 − 3.36

12. 22 − 14.989

13. 571 − 46.4

14. 4,225 − 391.0581

Subtracting a whole number from a decimal is a bit easier. Instead of having to borrow and rename over and over, we'll be subtracting 0 from a number over and over.

Example

3.34234 − 2

Line up the columns, place a decimal point after the subtrahend (2) and tack on five trailing zeros:

$$
\begin{array}{r}
3.34234 \\
-\ 2.00000 \\
\hline
1.34234
\end{array}
$$

Notice that the decimal part of our answer (the digits to the right of the decimal point) is the same as the decimal part of the minuend. That will always be true when we subtract a whole number from a number with a decimal part. In fact, you can bring the decimal part of the minuend right into your answer and just handle the subtraction of the whole number parts.

Example

329.8167 – 156

This time, subtract just the whole number parts:

$$\begin{array}{r} 329 \\ -\ \ 156 \\ \hline 173 \end{array}$$

Now, bring the decimal part of the minuend into the answer: 173.8167. The difference between 329.8167 and 156 is 173.8167.

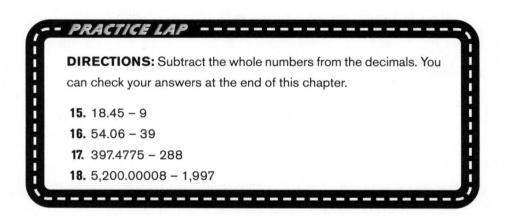

PRACTICE LAP

DIRECTIONS: Subtract the whole numbers from the decimals. You can check your answers at the end of this chapter.

15. 18.45 – 9

16. 54.06 – 39

17. 397.4775 – 288

18. 5,200.00008 – 1,997

SUBTRACTING DECIMALS FROM DECIMALS

The only real difference between subtracting decimals and subtracting whole numbers is the decimal point. Otherwise, it's the same subtraction we've been doing because grade school.

Example

5.64 – 2.87

Line up the decimal points. Because the minuend and the subtrahend have the same number of decimal places, no trailing zeros are needed:

$$
\begin{array}{r}
5.64 \\
-\ 2.87 \\
\hline
\end{array}
$$

Carry the decimal point down into the answer. Subtract, borrow and rename as needed:

$$
\begin{array}{r}
5.64 \\
-\ 2.87 \\
\hline
2.77
\end{array}
$$

Example

602.107 – 74.87

Line up the decimal points. Place a trailing zero on 74.87, subtract, and carry down the decimal point:

$$
\begin{array}{r}
602.107 \\
-\ 74.870 \\
\hline
527.237
\end{array}
$$

We saw how the sum of two decimals can have one fewer digit to the right of the decimal point when the sum of the digits in the rightmost column is 10. In the same way, when the difference is 0 between the digits in the rightmost column of a subtraction problem, the difference will have one fewer digit to the right of the decimal point than that of the minuend and subtrahend:

Example

84.236 – 26.876

Line up the decimal points, carry down the decimal point, and subtract:

$$
\begin{array}{r}
84.236 \\
-\ 26.876 \\
\hline
57.360
\end{array}
$$

The difference between 84.236 and 26.876 is 57.360, or 57.36.

IN EXAMPLES LIKE the preceding one, we can see why it is important to carry the decimal point down into your answer *before* subtracting. If you subtracted, discarded the trailing zero in the answer, and then placed the decimal point so that there were three places to its right, your answer would have been incorrect. Always carry the decimal point straight down, and discard trailing zeros after you have your answer. If this seems confusing, you can always keep trailing zeros on the end of your answers. An answer of 57.360 is correct.

PRACTICE LAP

DIRECTIONS: Subtract. You can check your answers at the end of this chapter.

19. 7.34 – 6.2
20. 38.766 – 9.558
21. 144.323 – 112.705
22. 89.0063 – 36.7223

MULTIPLYING DECIMALS BY WHOLE NUMBERS

We multiply a decimal by a whole number in a way similar to the way in which we multiply two whole numbers. We line up our columns, multiply, and carry, but when we are finished, our **product** has the same number of decimal places as the decimal **factor**.

FUEL FOR THOUGHT

IN A MULTIPLICATION problem, the two numbers being multiplied are called the **factors**, and the result of their multiplication is called the **product**. In the problem 4 × 3 = 12, 4 and 3 are factors and 12 is their product.

Example

1.2 × 6

Unlike with addition and subtraction, we do not line up the decimal points before multiplying. Instead, we write the problem vertically as we would any other multiplication problem.

$$
\begin{array}{r}
1.2 \\
\times\ 6 \\
\hline
\end{array}
$$

Before multiplying, we count the number of places to the right of the decimal point in the decimal factor. In this problem, the decimal factor 1.2 has one place to the right of the decimal point. That means that our answer will have one place to the right of the decimal point. Now we are ready to multiply:

$$
\begin{array}{r}
1.2 \\
\times\ 6 \\
\hline
7.2
\end{array}
$$

Let's look at another example.

Example

23.71 × 14

Write the problem vertically and count the number of digits to the right of the decimal point in the decimal factor. There are two digits to the right of the decimal point in 23.71. Our answer will have two digits to the right of the decimal point. You might even want to place the decimal point now before multiplying:

$$\begin{array}{r} 23.71 \\ \times\ \ 14\ . \\ \hline \end{array}$$

Now multiply:

$$\begin{array}{r} 23.71 \\ \times\ \ 14 \\ \hline 9\,4\,8\,4 \\ +\ 2\,3\,7\,1 \\ \hline 3\,3\,1.94 \end{array}$$

The product of 23.71 and 14 is 331.94.

CAUTION!

AS WITH ADDITION and subtraction, if the last digit of your answer, to the right of the decimal point is zero, then that trailing zero can be removed from your answer. But as with addition and subtraction, be sure to place the decimal point in the correct place before removing any trailing zeros.

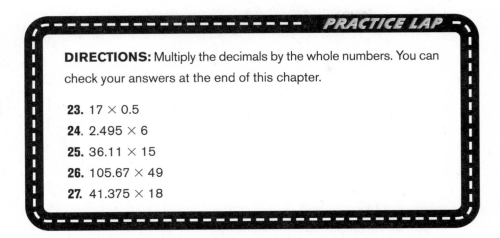

DIRECTIONS: Multiply the decimals by the whole numbers. You can check your answers at the end of this chapter.

23. 17 × 0.5

24. 2.495 × 6

25. 36.11 × 15

26. 105.67 × 49

27. 41.375 × 18

MULTIPLYING DECIMALS BY DECIMALS

When both factors being multiplied are decimals, we must add the number of digits to the right of the decimal point in the first factor to the number of digits to the right of the decimal point in the second factor. That sum is the number of digits to the right of the decimal point in our product.

When we multiplied a decimal number by a whole number, the number of digits to the right of the decimal point of our answer was equal to the number of digits to the right of the decimal point in our decimal number. That's because the number of digits to the right of the decimal point of the whole number was zero.

So, now we have a rule we can use whether we are multiplying a decimal by a decimal or a decimal by a whole number: Add the number of digits to the right of the decimal point in the first factor to the number of digits to the right of the decimal point in the second factor. The sum is the number of digits to the right of the decimal point of the product.

Example

5.6 × 2.7

The first factor has one digit to the right of the decimal point and the second factor has one digit to the right of the decimal point. Therefore, our product will have two digits to the right of the decimal point:

$$
\begin{array}{r}
5.6 \\
\times\ 2.7 \\
\hline
392 \\
+\ 112 \\
\hline
15.12
\end{array}
$$

Let's try one more:

Example

1.2534×24.05

The first factor has four digits to the right of the decimal point and the second factor has two digits to the right of the decimal point. Our product will have six digits to the right of the decimal point:

$$
\begin{array}{r}
1.2534 \\
\times\ 24.05 \\
\hline
626\ 70 \\
0 \\
5\,0136 \\
+\ 2\,5068 \\
\hline
30.144\,2\,7\,0
\end{array}
$$

Now that we've placed our decimal point with six digits to its right, we can remove the trailing zero. The product of 1.2534 and 24.05 is 30.14427.

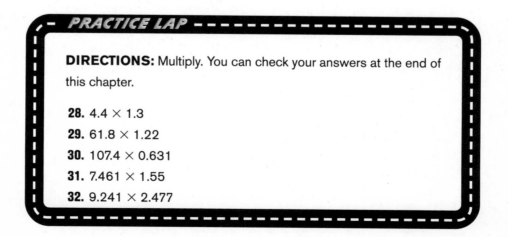

PRACTICE LAP

DIRECTIONS: Multiply. You can check your answers at the end of this chapter.

28. 4.4×1.3

29. 61.8×1.22

30. 107.4×0.631

31. 7.461×1.55

32. 9.241×2.477

DIVIDING DECIMALS BY WHOLE NUMBERS

Dividing a decimal by a whole number is just like dividing a whole number by a whole number. The number of digits to the right of the decimal point in your answer is equal to the number of digits to the right of the decimal point of the **dividend**.

FUEL FOR THOUGHT

IN A DIVISION problem, the number being divided is called the **dividend**. The number by which the dividend is divided is called the **divisor**. The result of the division is called the **quotient**. In the problem $12 \div 3 = 4$, 12 is the dividend, 3 is the divisor, and 4 is the quotient.

Example

$4.5 \div 9$

The dividend has one digit to the right of the decimal point, so our answer will have one digit to the right of the decimal point.

$$\begin{array}{r} 0.5 \\ 9\overline{)4.5} \end{array}$$

Just as when we divide a whole number by a whole number, there is often a remainder. When we divide a whole number by a whole number and there is a remainder, we add a decimal point to the dividend followed by a zero. We continue adding zeros until we are done dividing. When the dividend already has a decimal point, we just add zeros until we are done dividing.

Example

59.13 ÷ 4

The dividend has two digits to the right of the decimal point, so our answer will have two digits to the right of the decimal point.

```
        14.78
   4 )59.13
       4
      ‾‾‾
      19
      16
      ‾‾‾
       31
       28
      ‾‾‾
        33
        32
       ‾‾‾
         1
```

We have one left over. Now, we add a zero to the dividend and continue dividing. Our answer will have more than two digits to the right of the decimal point—but now our dividend has more than two digits to the right of the decimal point.

```
        14.782
   4 )59.130
       4
      ‾‾‾
      19
      16
      ‾‾‾
       31
       28
      ‾‾‾
        33
        32
       ‾‾‾
         10
          8
        ‾‾‾
          2
```

Now we have two left over, so we add another zero to the dividend and continue dividing.

```
        14.7825
    4)59.1300
      4
      19
      16
       31
       28
        33
        32
         10
          8
         20
         20
          0
```

Finally, we have no remainder. We began with a dividend that had two digits to the right of the decimal point, and we expected our answer to have two digits to the right of the decimal point. In the end, our dividend had four digits to the right of the decimal point, and our answer had four digits to the right of the decimal point. The rule holds true: *The number of digits to the right of the decimal point in the answer is equal to the number of digits to the right of the decimal point of the dividend.* We just don't know how many digits that is until we're done dividing!

Some division problems could go on forever. Quotients like that can be non-terminating decimals or they can be repeating decimals. We'll learn more about those in Chapter 11. For now, we'll round our decimals to four places. Still remember how to round decimals?

Example

16.37 ÷ 3

Because the dividend has two digits to the right of the decimal point, we will expect our answer to have two digits to the right of the decimal point—but we will see if that number goes up.

$$
\begin{array}{r}
5.45 \\
3\overline{)16.37} \\
\underline{15} \\
13 \\
\underline{12} \\
17 \\
\underline{15} \\
2
\end{array}
$$

Now we have two left over, so we add a zero to the dividend and continue dividing.

$$
\begin{array}{r}
5.456 \\
3\overline{)16.370} \\
\underline{15} \\
13 \\
\underline{12} \\
17 \\
\underline{5} \\
20 \\
\underline{18} \\
2
\end{array}
$$

Again, we have two left over. We'll add another zero.

$$
\begin{array}{r}
5.4566 \\
3\overline{)16.3700} \\
\underline{15} \\
13 \\
\underline{12} \\
17 \\
\underline{15} \\
20 \\
\underline{18} \\
20 \\
\underline{18} \\
2
\end{array}
$$

Another two left over. No matter how many zeros we had, we will always add another 6 onto the end of our quotient and have two left over. We will round our decimal to four places. In other words, we will round to the nearest ten thousandth. 5.45666 has a 6 in the hundred thousandths place, so we must round up. 5.45666 to the nearest ten thousandth is 5.4567.

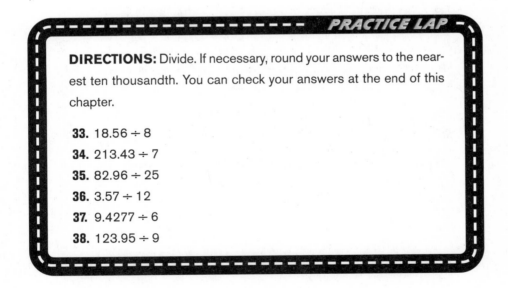

DIRECTIONS: Divide. If necessary, round your answers to the nearest ten thousandth. You can check your answers at the end of this chapter.

33. 18.56 ÷ 8

34. 213.43 ÷ 7

35. 82.96 ÷ 25

36. 3.57 ÷ 12

37. 9.4277 ÷ 6

38. 123.95 ÷ 9

DIVIDING DECIMALS BY DECIMALS

There is one major difference between dividing decimals by whole numbers and dividing decimals by decimals. Before you can divide a decimal by a decimal, you must convert the divisor into a whole number. To do this, you must shift the decimal point to the right. In order to keep value of the equation the same, every time you shift the decimal point of the divisor, you must also shift the decimal point of the dividend.

Example

$8.6 \div 4.3$

Think back to fractions. We saw how the value of a fraction did not change if we multiplied its numerator and its denominator by the same number. A division problem is just like a fraction—remember, a fraction *is* division. When we say "$8.6 \div 4.3$," that is the same as $\frac{8.6}{4.3}$. When we convert a fraction to a number with a different denominator, we are multiplying the dividend and divisor by the same number.

Now let's go back to decimals. Shifting a decimal point to the right is the same as multiplying a number by 10. Instead of dividing 8.6 by 4.3, we are going to divide 86 by 43. We just shifted the decimal point of the divisor, 4.3, one place to the right, and we shifted the decimal point of the dividend, 8.6, one place to the right. This is the same as if we had multiplied the fraction $\frac{8.6}{4.3}$ by $\frac{10}{10}$. The value of the fraction, or division problem, hasn't changed at all. Now we are dividing a whole number by a whole number—we know how to do that!

$$\begin{array}{r} 2 \\ 43\overline{)86} \end{array}$$

In this problem, division of two decimals gave us a quotient that had no places to the right of the decimal point. But that is no surprise: look at the dividend. Our final dividend had no places to the right of the decimal point, and our answer has no places to the right of the decimal point.

Example

$8 \div 1.6$

In this example, our divisor has one digit to the right of the decimal point, so we must shift the decimal point one place. The dividend has no places to the right of the decimal point. We must add a decimal point to the dividend, and then shift it to the right. Remember, shifting a decimal point to the right is the same as multiplying by 10: $8 \div 1.6$ becomes $80 \div 16$. We're back to dividing whole numbers.

$$\begin{array}{r} 5 \\ 16\overline{)80} \end{array}$$

Not all decimal division problems turn into whole number division problems. Sometimes the dividend has more digits to the right of the decimal point than the divisor has such digits. And just like we saw when we divided decimals by whole numbers, sometimes we need to add digits to the dividend.

Example

62.314 ÷ 12.5

The divisor has one digit to the right of the decimal point. Shift the decimal point in both the divisor and the dividend one place to the right: 623.14 ÷ 125. The problem has become a decimal divided by a whole number—and we just learned how to solve that kind of problem.

$$
\begin{array}{r}
4.98512 \\
125\overline{)623.14000} \\
\underline{500} \\
1231 \\
\underline{1125} \\
1064 \\
\underline{1000} \\
64 \\
\underline{0} \\
640 \\
\underline{625} \\
150 \\
\underline{125} \\
250 \\
\underline{250} \\
0
\end{array}
$$

We could round this answer to the ten thousandths place, but 4.98512 is an exact answer.

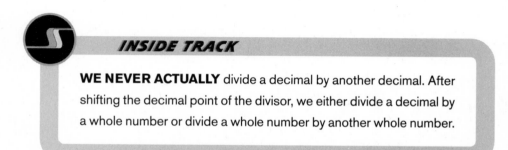

INSIDE TRACK

WE NEVER ACTUALLY divide a decimal by another decimal. After shifting the decimal point of the divisor, we either divide a decimal by a whole number or divide a whole number by another whole number.

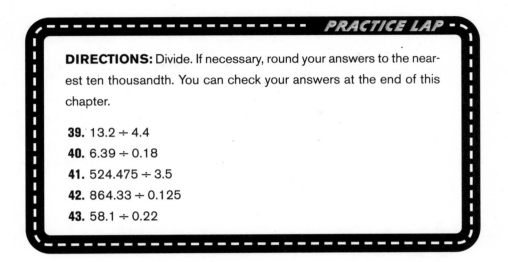

PRACTICE LAP

DIRECTIONS: Divide. If necessary, round your answers to the nearest ten thousandth. You can check your answers at the end of this chapter.

39. 13.2 ÷ 4.4
40. 6.39 ÷ 0.18
41. 524.475 ÷ 3.5
42. 864.33 ÷ 0.125
43. 58.1 ÷ 0.22

Now you can add, subtract, multiply, and divide with decimals, and perform those same operations with a mix of decimals and whole numbers. You're ready to move on to percents.

ANSWERS

1. 56.35 (two digits to the right of the decimal point in 2.35, two digits to the right of the decimal point in the answer)

2. 30.568 (three digits to the right of the decimal point in 27.568, three digits to the right of the decimal point in the answer)

3. 289.051 (three digits to the right of the decimal point in 0.051, three digits to the right of the decimal point in the answer)

4. 1,119.65722 (five digits to the right of the decimal point in 74.65722, five digits to the right of the decimal point in the answer)

5. 882.74 (two digits to the right of the decimal point in 8.74, two digits to the right of the decimal point in the answer)

6. 86.26

7. 116.87

8. 6,335.61

9. 226.599

10. 121.2966

11. 5.64 (two digits to the right of the decimal point in 3.36, two digits to the right of the decimal point in the answer)

12. 7.011 (three digits to the right of the decimal point in 14.989, three digits to the right of the decimal point in the answer)

13. 524.6 (one digit to the right of the decimal point in 46.4, one digit to the right of the decimal point in the answer)

14. 3,833.9419 (four digits to the right of the decimal point in 391.0581, four digits to the right of the decimal point in the answer)

15. 9.45

16. 15.06

17. 109.4775

18. 3,203.00008

19. 1.14

20. 29.208

21. 31.618

22. 52.2840, or 52.284

23. The decimal factor has one digit to the right of the decimal point, so our answer will have one digit to the right of the decimal point:

$$\begin{array}{r} 1\,7 \\ \times\,0.5 \\ \hline 8.5 \end{array}$$

24. The decimal factor has three digits to the right of the decimal point, so our answer will have three digits to the right of the decimal point:

$$\begin{array}{r} 2.495 \\ \times\,6 \\ \hline 14.970,\ \text{or}\ 14.97 \end{array}$$

25. The decimal factor has two digits to the right of the decimal point, so our answer will have two digits to the right of the decimal point:

$$\begin{array}{r} 36.11 \\ \times\,15 \\ \hline 1805\,5 \\ +\ \ 3611 \\ \hline 54\,1.65 \end{array}$$

26. The decimal factor has two digits to the right of the decimal point, so our answer will have two digits to the right of the decimal point:

$$
\begin{array}{r}
105.67 \\
\times\ \ 49 \\
\hline
95103 \\
+\,42268\ \ \\
\hline
5{,}177.83
\end{array}
$$

27. The decimal factor has three digits to the right of the decimal point, so our answer will have three digits to the right of the decimal point:

$$
\begin{array}{r}
41.375 \\
\times\ \ \ 18 \\
\hline
331000 \\
+\,41375\ \ \\
\hline
744750, \text{ or } 744.75
\end{array}
$$

28. 4.4 has one digit to the right of the decimal point and 1.3 has one digit to the right of the decimal point, so our answer has two digits to the right of the decimal point: 5.72.

29. 61.8 has one digit to the right of the decimal point and 1.22 has two digits to the right of the decimal point, so our answer has three digits to the right of the decimal point: 75.396.

30. 107.4 has one digit to the right of the decimal point and 0.631 has three digits to the right of the decimal point, so our answer has four digits to the right of the decimal point: 67.7694.

31. 7.461 has three digits to the right of the decimal point and 1.55 has two digits to the right of the decimal point, so our answer has five digits to the right of the decimal point: 11.56455.

32. 9.241 has three digits to the right of the decimal point and 2.477 has three digits to the right of the decimal point, so our answer has six digits to the right of the decimal point: 22.889957.

33.

$$
\begin{array}{r}
2.32 \\
8\overline{)18.56} \\
\underline{16} \\
25 \\
\underline{24} \\
16 \\
\underline{16} \\
0
\end{array}
$$

34.

$$
\begin{array}{r}
30.49 \\
7\overline{)213.43} \\
\underline{21} \\
03 \\
\underline{0} \\
34 \\
\underline{28} \\
63 \\
\underline{63} \\
0
\end{array}
$$

35. Add two zeros to the dividend:

$$
\begin{array}{r}
3.3184 \\
25\overline{)82.9600} \\
\underline{75} \\
79 \\
\underline{75} \\
46 \\
\underline{25} \\
210 \\
\underline{200} \\
100 \\
\underline{100} \\
0
\end{array}
$$

36. Add two zeros to the dividend:

$$
\begin{array}{r}
0.2975 \\
12\overline{)3.5700} \\
\underline{0} \\
35 \\
\underline{24} \\
117 \\
\underline{108} \\
90 \\
\underline{84} \\
60 \\
\underline{60} \\
0
\end{array}
$$

37. Add one zero to the dividend and round. Because the hundred thousandths digit of 1.57128 is an 8, 1.57128 rounds up to 1.5713.

$$
\begin{array}{r}
1.57128 \\
6\overline{)9.42770} \\
\underline{6} \\
34 \\
\underline{30} \\
42 \\
\underline{42} \\
07 \\
\underline{6} \\
17 \\
\underline{12} \\
50 \\
\underline{48} \\
2
\end{array}
$$

38. Add three zeros to the dividend and round. Because the hundred thousandths digit of 13.77222 is a 2, 13.77222 rounds down to 13.7722.

$$
\begin{array}{r}
13.77222 \\
9\overline{)123.95000} \\
9 \\ \hline
33 \\
27 \\ \hline
69 \\
63 \\ \hline
65 \\
63 \\ \hline
20 \\
18 \\ \hline
20 \\
18 \\ \hline
20 \\
18 \\ \hline
2
\end{array}
$$

39. Shift the decimal point of both the divisor and the dividend one place to the right. $13.2 \div 4.4$ becomes $132 \div 44$:

$$
\begin{array}{r}
3 \\
44\overline{)132} \\
132 \\ \hline
0
\end{array}
$$

40. Shift the decimal point of both the divisor and the dividend two places to the right. 6.39 ÷ 0.18 becomes 639 ÷ 18:

$$
\begin{array}{r}
35.5 \\
18\overline{)639.0} \\
\underline{54} \\
99 \\
\underline{90} \\
90 \\
\underline{90} \\
0
\end{array}
$$

41. Shift the decimal point of both the divisor and the dividend one place to the right. 524.475 ÷ 3.5 becomes 5,244.75 ÷ 35:

$$
\begin{array}{r}
149.85 \\
35\overline{)5244.75} \\
\underline{35} \\
174 \\
\underline{140} \\
344 \\
\underline{315} \\
297 \\
\underline{280} \\
175 \\
\underline{175} \\
0
\end{array}
$$

42. Shift the decimal point of both the divisor and the dividend three places to the right. 864.33 ÷ .125 becomes 864,330 ÷ 125:

$$
\begin{array}{r}
6914.64 \\
125\overline{)864330.00} \\
750 \\
\overline{1143} \\
1125 \\
\overline{183} \\
125 \\
\overline{580} \\
500 \\
\overline{800} \\
750 \\
\overline{500} \\
500 \\
\overline{0}
\end{array}
$$

43. Shift the decimal point of both the divisor and the dividend two places to the right. 58.1 ÷ 0.22 becomes 5,810 ÷ 22:

$$
\begin{array}{r}
264.09090 \\
22\overline{)5810.00000} \\
44 \\
\hline
141 \\
132 \\
\hline
90 \\
88 \\
\hline
20 \\
0 \\
\hline
200 \\
198 \\
\hline
20 \\
0 \\
\hline
200 \\
198 \\
\hline
20 \\
0 \\
\hline
20
\end{array}
$$

Because the hundred thousandths digit of 264.09090 is a zero, 264.09090 rounds down to 264.0909.

What's a Percent?

WHAT'S AROUND THE BEND

- ➡ Understanding Percents
- ➡ Writing Percents as Decimals
- ➡ Finding a Percent of a Number
- ➡ Representing a Part of a Number as a Percent
- ➡ Percents Greater Than 100
- ➡ Percents Less Than 1

In this chapter, we'll learn about percents: what they are and how to find them.

IT'S OUT OF A HUNDRED

A **percent** is a number out of 100. We use the symbol % to represent the word *percent*. The value 36% is read as "36 percent" and represents 36 out of 100.

> ### FUEL FOR THOUGHT
>
> **A PERCENT IS** a ratio that represents a part-to-a-whole as a number out of 100. Fractions are used to represent a part-to-a-whole, but fractions can have many different denominators. A percent is like a fraction whose denominator is always 100. The relationship "50 out of 100" is written as 50%. The relationship "1 out of 2" can also be written as 50%. That is because the fraction $\frac{1}{2}$ is equal to $\frac{50}{100}$. Fractions and decimals can be written as percents, and vice versa, as we'll see in Chapter 11.

Example

67% of Tim's marbles are red. What does this mean?

67% represents "67 out of 100." For every 100 marbles Tim has, 67 of them are red.

Example

28 of every 100 pizzas that Manny makes are pepperoni pizzas. What percent of Manny's pizzas are pepperoni pizzas?

28 out of 100 is 28%. If Manny made 100 pizzas, 28 of them would be pepperoni—but what if Manny made only 75 pizzas? Before we can answer this question, we must learn how to write percents as decimals.

WRITING PERCENTS AS DECIMALS

We know now that a percent is a number out of 100 We can say that 28% is 28 out of 100, or 28 hundredths. How do we write 28 hundredths as a decimal? 0.28. Because a percent is a number out of 100, we can rewrite a percent as a decimal by removing the percent symbol and dividing the number by 100. This causes the decimal to move two places to the left: 28% = 0.28.

Example

What is 43% as a decimal?

You know that 43% is 43 out of 100, or 43 hundredths. Moving the decimal point two places to the left (or dividing by 100), 43% becomes 0.43.

Example

What is 79% as a decimal?

Move the decimal point two places to the left: 79% = 0.79.

CAUTION!

IF YOU MOVE the decimal point two places to the left of a single-digit number, you will first have to add a zero to the left of the number. For example, to write 2% as a decimal, we must add a leading zero to 2%: 02%. Now we can move the decimal point two places to the left, and 2% becomes .02.

PRACTICE LAP

DIRECTIONS: Write each percent as a decimal. You can check your answers at the end of this chapter.

1. 24%

2. 80%

3. 11%

4. 97%

5. 6%

Now that we know how to write percents as decimals, we can go back to Manny's pizzeria.

FINDING THE PERCENT OF A NUMBER

If 28% of Manny's pizzas are pepperoni pizzas, and Manny makes 75 pizzas, how many pepperoni pizzas does Manny make? In other words, what is 28% *of* 75? To find the percent of a number, first, write the percent as a decimal. Then, multiply the decimal by the number—good thing we already know how to multiply decimals! 28% = 0.28. 0.28 × 75 = 21. If Manny makes 75 pizzas, 21 of them are pepperoni pizzas.

Example

What is 75% of 64?

First, write 75% as a decimal: 75% = 0.75. Then, multiply the decimal by the number: 0.75 × 64 = 48.

Example

Jason takes 60 photos. If 15% of them are black and white, how many black-and-white photos does he take?

This problem is asking us to find 15% of 60. Write 15% as a decimal: 15% = 0.15. Multiply the decimal by 60: 0.15 × 60 = 9. Jason took 9 black-and-white photos.

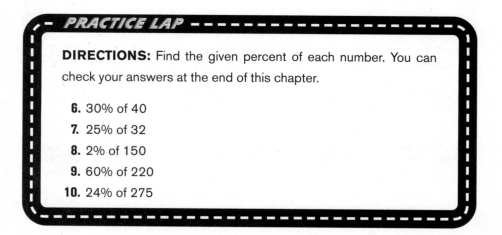

PRACTICE LAP

DIRECTIONS: Find the given percent of each number. You can check your answers at the end of this chapter.

6. 30% of 40

7. 25% of 32

8. 2% of 150

9. 60% of 220

10. 24% of 275

Sometimes, a percent of a whole number results in a decimal.

Example

What is 23% of 54?

23% = 0.23, 0.23 × 54 = 12.42; 23% of 54 is 12.42.

Percents themselves might be written as decimals. We follow the same steps to find the decimal percent of a number.

Example

What is 5.2% of 90?

Begin as always by writing the percent as a decimal: shift the decimal point two places to the left: 5.2% = 0.052. Now multiply: 0.052 × 90 = 4.68.

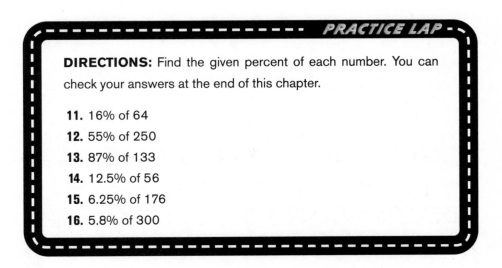

PRACTICE LAP

DIRECTIONS: Find the given percent of each number. You can check your answers at the end of this chapter.

11. 16% of 64

12. 55% of 250

13. 87% of 133

14. 12.5% of 56

15. 6.25% of 176

16. 5.8% of 300

REPRESENTING A PART OF A NUMBER AS A PERCENT

Let's look again at the last example. Instead of saying that 5.2% of 90 is 4.68, we could say that 4.68 is 5.2% of 90. But if we didn't already know that, how could we find what percent of 90 is 4.68? To find the number that is 5.2% of 90, we multiplied 90 by 0.052. To find what percent 4.68 is of 90, we divide 4.68 by 90. However, when we finish dividing, we have a quotient of .052. How do we write a decimal as a percent? By moving the decimal point two places to the right and adding the percent sign: .052 = 5.2%.

Writing Decimals as Percents

To write a percent as a decimal, we moved the decimal point two places to the left. So, it makes sense that in order to write a decimal as a percent, we move the decimal point two places to the right. In this case, you are multiplying the number by 100.

Example

Start with 42%. Drop the percent sign and move the decimal point two places to the left. 42% becomes 0.42. Now switch back: move the decimal point two places to the right and add the percent sign: 0.42 becomes 42% again.

Example

Write 0.72 as a percent.

Move the decimal point two places to the right and add the percent sign: 0.72 = 72%.

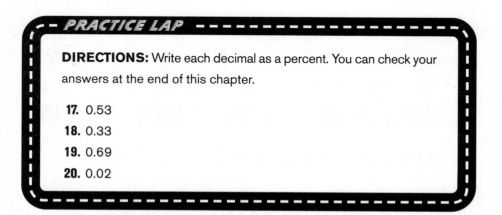

PRACTICE LAP

DIRECTIONS: Write each decimal as a percent. You can check your answers at the end of this chapter.

17. 0.53
18. 0.33
19. 0.69
20. 0.02

Now that we are comfortable writing decimals as percents, let's represent a part of a number as a percent.

Example

What percent is 12 of 48?

Divide 12 by 48: 12 ÷ 48 = 0.25. To write 0.25 as a percent, move the decimal point two places to the right and add the percent sign: 0.25 = 25%.

As we saw earlier, percents themselves can be decimals.

Example

30 is what percent of 96?

Divide 30 by 96: 30 ÷ 96 = 0.3125. Move the decimal point two places to the right and add the percent sign: 0.3125 = 31.25%.

Sometimes, a part of a number cannot be represented as an exact decimal or percent. When this happens, we must round.

Example

15 is what percent of 46?

Divide 15 by 46: 15 ÷ 46 = 0.3261, rounded to the nearest ten thousandth. Move the decimal point two places to the right and add the percent sign: 0.3261 = 32.61%.

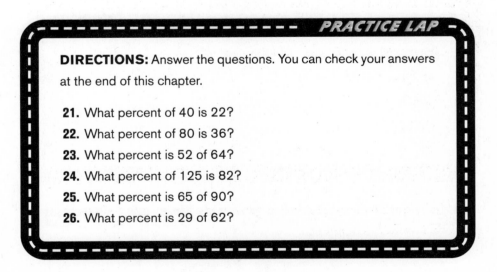

PRACTICE LAP

DIRECTIONS: Answer the questions. You can check your answers at the end of this chapter.

21. What percent of 40 is 22?
22. What percent of 80 is 36?
23. What percent is 52 of 64?
24. What percent of 125 is 82?
25. What percent is 65 of 90?
26. What percent is 29 of 62?

We've found the percent of a number, and we've found what percent a new number is of an original number. Now, given the new number and the percent, let's find the original number.

Example

15 is 10% of what number?

In other words, 15, our new number, is what we have after taking 10% of an original number. We can find the original number by working backward. First, write 10% as a decimal. Move the decimal point two places to the left and remove the percent sign: 10% = 0.10.

We know that 10% of a number can be found by multiplying that number by 0.10. Because we have the new number, we can divide the new number by the percent to find the original number: 15 ÷ 0.10 = 150. The original number is 150: 15 is 10% of 150.

Example

26.88 is 32% of what number?

Convert 32% to a decimal and divide the new number, 26.88, by that decimal: 32% = 0.32, 26.88 ÷ 0.32 = 84. The original number is 84. 26.88 is 32% of 84.

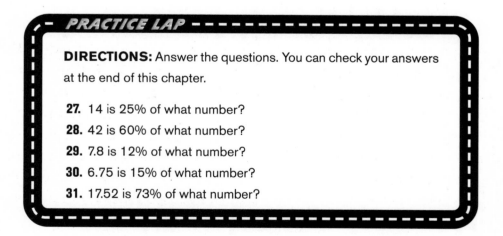

PRACTICE LAP

DIRECTIONS: Answer the questions. You can check your answers at the end of this chapter.

27. 14 is 25% of what number?

28. 42 is 60% of what number?

29. 7.8 is 12% of what number?

30. 6.75 is 15% of what number?

31. 17.52 is 73% of what number?

THE EASIEST PERCENT with which to work is 100%. 100% of a quantity is the entire quantity. What is 100% of 10? 10. 100% of 600 is 600, 100% of 5.4241 is 5.4241, and 100% of 1,000,000 is 1,000,000. 100% as a decimal is the number 1. Why is that? Think about how a percent is written as a decimal: by moving the decimal point two places to the left. If you have all of a quantity, then you have 100% of the quantity. What percent of 65 is 65? 100%; 262 is 100% of 262; 85.201 is 100% of 85.201; and 1,000,000 is 100% of 1,000,000.

LOOK AT WHAT page of this book you are on right now, and look at the total number of pages in this book. What percent of the book have you completed so far? As you continue reading, after every ten pages, figure out what percent of the book you have read, and what percent of the book remains to be read.

PERCENTS GREATER THAN 100

A percent is a number out of 100, but a percent can be greater than 100. A percent greater than 100 means that you don't have part of a whole, you have a quantity that is greater than the whole. That might sound impossible, because in real life, you cannot have more than 100% of a real quantity. For instance, you can have 100% of a pie, but you can't have 120% of the pie. However, when we're working with numbers, we need to be able to answer the question, "What percent of 4 is 6?" or "What is 120% of 10?" But here's the good news: We work with percents that are greater than 100 in the same way we work with percents that are less than 100.

Example

What is 120% of 85?

We follow the same steps we used when working with percents that are less than 100. Write the percent as a decimal and multiply. Move the decimal point two places to the left: $120\% = 1.20$ or 1.2: $1.2 \times 85 = 102$. Whenever we find a percent greater than 100 of a quantity, the answer will be greater than the original quantity. This is because a percent greater than 100 represents a value greater than the whole.

Example

What is 200% of 23?

$200\% = 2.00$, or simply 2: $2 \times 23 = 46$. If we have 200% of a quantity, then we have twice that quantity. If we have 300% of a quantity, then we have three times that quantity.

You might be thinking that there is no reason to learn about percents that are greater than 100% if they can never happen in real life. Well, here's a real-life example:

Example

Jordan travels 24 miles to a wedding. Bria travels 150% farther than Jordan to attend the same wedding. How far does Bria travel?

We must find the quantity that is 150% of 24.
$150\% = 1.5$, $1.5 \times 24 = 36$. Bria travels 36 miles to the wedding.

Example

What percent of 5 is 8?

Even though 8 is greater than 5, we still follow the same steps to find the percent of 5 that 8 represents. $8 \div 5 = 1.6$. Write the decimal as a percent by moving the decimal point two places to the right. $1.6 = 160\%$; 8 is 160% of 5.

Example

Marie has 80 books and Al has 64 books. What percent of the size of Al's book collection is the size of Marie's book collection?

In other words, what percent of 64 is 80? $80 \div 64 = 1.25 = 125\%$

CAUTION!

IF YOU ARE working with two different numbers, the percent that the first number is of the second number is not equal to the percent that the second number is of the first number. That sounds confusing, so let's look at some actual numbers. What percent of 10 is 5? $5 \div 10 = 0.5$, or 50%. 5 is 50% of 10. What percent of 5 is 10? $10 \div 5 = 2$, or 200%. The percent that 5 is of 10 (50%) is not the same percent that 10 is of 5 (200%). Be careful when you read a question: Unless the two numbers are the same, the order of the numbers matters.

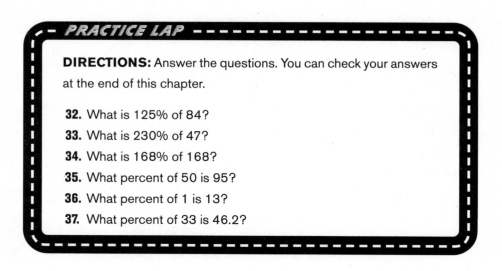

PRACTICE LAP

DIRECTIONS: Answer the questions. You can check your answers at the end of this chapter.

32. What is 125% of 84?

33. What is 230% of 47?

34. What is 168% of 168?

35. What percent of 50 is 95?

36. What percent of 1 is 13?

37. What percent of 33 is 46.2?

PERCENTS LESS THAN 1

Just as we have percents that are greater than 100, we can also have percents that are less than 1. Sometimes, a part of a whole is so small that it represents less than 1% of the whole. That's not a problem. We've already seen that percents can be decimals.

Example

What is 0.2% of 22?

We follow the same steps we used when working with percents that are greater than 1. Write the percent as a decimal and multiply. Move the decimal point two places to the left: $0.2\% = 0.002$. $0.002 \times 22 = 0.044$. When we find less than 1% of a quantity, our answer will be much, much smaller than the original quantity.

Example

What is 0.08% of 600?

Watch the decimal point carefully. $0.08\% = 0.0008$. $0.0008 \times 600 = 0.48$.

Example

What percent of 250 is 1?

Divide 1 by 250: $1 \div 250 = 0.004$. Move the decimal point two places to the right. $0.004 = 0.4\%$.

Example

Alex must lay 625 bricks to make a wall. If he has laid 5 bricks so far, what percent of the wall is complete?

In other words, what percent of 625 is 5?
$5 \div 625 = 0.008 = 0.8\%$

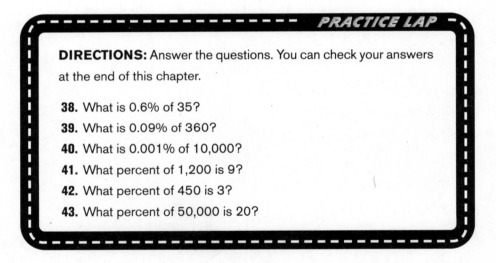

PRACTICE LAP

DIRECTIONS: Answer the questions. You can check your answers at the end of this chapter.

38. What is 0.6% of 35?

39. What is 0.09% of 360?

40. What is 0.001% of 10,000?

41. What percent of 1,200 is 9?

42. What percent of 450 is 3?

43. What percent of 50,000 is 20?

Working with percents is a lot like working with decimals, because we always convert percents to decimals before performing any operations with them. Next, we'll look at more complex operations with percents.

ANSWERS

1. $24\% = 0.24$

2. $80\% = 0.80$, or 0.8

3. $11\% = 0.11$

4. $97\% = 0.97$

5. $6\% = 0.06$

6. $30\% = 0.3$, $0.3 \times 40 = 12$

7. $25\% = 0.25$, $0.25 \times 32 = 8$

8. $2\% = 0.02$, $0.02 \times 150 = 3$

9. $60\% = 0.6$, $0.6 \times 220 = 132$

10. $24\% = 0.24$, $0.24 \times 275 = 66$

11. $16\% = 0.16$, $0.16 \times 64 = 10.24$

12. $55\% = 0.55$, $0.55 \times 250 = 137.5$

13. $87\% = 0.87$, $0.87 \times 133 = 115.71$

14. $12.5\% = 0.125$, $0.125 \times 56 = 7$

15. $6.25\% = 0.0625$, $0.0625 \times 176 = 11$

16. $5.8\% = 0.058$, $0.058 \times 300 = 17.4$

17. $0.53 = 53\%$

18. $0.33 = 33\%$

19. $0.69 = 69\%$

20. $0.02 = 2\%$

21. $22 \div 40 = 0.55 = 55\%$

22. $36 \div 80 = 0.45 = 45\%$

23. $52 \div 64 = 0.8125 = 81.25\%$

24. $82 \div 125 = 0.656 = 65.6\%$

25. $65 \div 90 = 0.7222$ to the nearest ten thousandth. $0.7222 = 72.22\%$

26. $29 \div 62 = 0.4677$ to the nearest ten thousandth. $0.4677 = 46.77\%$

27. $25\% = 0.25$, $14 \div 0.25 = 56$

28. $60\% = 0.60$, $42 \div 0.60 = 70$

29. 12% = 0.12, 7.8 ÷ 0.12 = 65

30. 15% = 0.15, 6.75 ÷ 0.15 = 45

31. 73% = 0.73, 17.52 ÷ 0.73 = 24

32. 125% = 1.25, 84 ÷ 1.25 = 105

33. 230% = 2.3, 47 ÷ 2.3 = 108.1

34. 168% = 1.68, 168 ÷ 1.68 = 282.24

35. 95 ÷ 50 = 1.9 = 190%

36. 13 ÷ 1 = 13 = 1,300%

37. 46.2 ÷ 33 = 1.4 = 140%

38. 0.6% = 0.006, 35 × 0.006 = 0.21

39. 0.09% = 0.0009, 360 × 0.0009 = 0.324

40. 0.001% = 0.00001, 10,000 × 0.00001 = 0.1

41. 9 ÷ 1,200 = 0.0075 = 0.75%

42. 3 ÷ 450 = 0.0067 to the nearest ten thousandth. 0.0067 = 0.67%

43. 20 ÷ 50,000 = 0.0004 = 0.04%

10

Using Percents

When comparing a change in data, we often want to make a statement about the significance of the change. In this chapter, we'll learn how to find the percent increase or decrease from one value to another, and how to use percents to solve real-life, two-step problems.

PERCENT INCREASE

The Dayley High School football team won eight games last season and 12 games this season. Subtraction can tell us that the team won four more games in this season, but by what percent did the team's win total increase? Sometimes, a **percent increase** can tell us more about a change in values than subtraction can tell us.

FUEL FOR THOUGHT

PERCENT INCREASE IS the difference between an original value and a new value divided by the original value. When finding a percent increase, the original value is subtracted from the new value, because the new value is larger than the original value. Percent increase is used to show the growth from an original value to a new value.

Why can a percent increase be a more interesting statistic than simply subtracting the new value from the old value? Percent increase can provide context to a situation. Just using subtraction gives us a difference, but it can be hard to tell how significant that difference is.

The Dayley High School football team increased its win total by four. Let's say a machine produces 1,500 paper clips in its first hour and 1,504 paper clips in its second hour. That's also an increase of four, because $1,504 - 1,500 = 4$. But which change was more significant? Let's find the percent increase of each statistic.

To find the percent increase from an original value to a new value, begin by subtracting the original value from the new value. Next, divide that difference by the original value. This will give you a decimal number. Write that decimal as a percent by moving the decimal point two places to the right.

Example

The Dayley High School football team won 8 games last season and 12 games this season.

In this problem, the original value is 8 and the new value is 12. Subtract the original value from the new value: $12 - 8 = 4$. Next, divide the difference by the original value: $4 \div 8 = 0.5$. Now, write the decimal as a percent by moving the decimal point two places to the right: $0.5 = 50\%$. The Dayley High School football team increased its win total by 50% from last season to this season.

Example

A machine produces 1,500 paper clips in its first hour and 1,504 paper clips in its second hour.

Follow the same steps. Subtract the original value from the new value: 1,504 − 1,500 = 4. Divide the difference by the original value: 4 ÷ 1,500 = .0027, to the nearest ten thousandth. Move the decimal point two places to the right: .0027 = 0.27%.

The football team and the paper clip producing machine both increased their respective totals by four, but the football team's increase was much more significant. Relative to their first season, the team increased its win total by 50%, while the paper clip producing machine only increased its production total by a small 0.27%. It might be hard to even notice the increase made by the paper clip machine, but the increase made by the football team was very significant.

INSIDE TRACK

WHEN THE ORIGINAL and new values that are being compared are small, even a small change can represent a large percent increase. When the original and new values that are being compared are large, even a large change might not represent a large percent increase. For the paper clip producing machine to experience a growth as large as the football team's growth (50%), it would have had to increase its production from 1,500 paper clips in an hour to 2,250 paper clips in an hour!

Sometimes, the new value can be so much larger than the old value that the percent increase is 100%, or even higher.

Example

Find the percent increase from 12 to 30.

The new value is more than double the old value, but we still follow the same steps. Subtract the original value from the new value and divide by the original value: 30 – 12 = 18, 18 ÷ 12 = 1.5 = 150%.

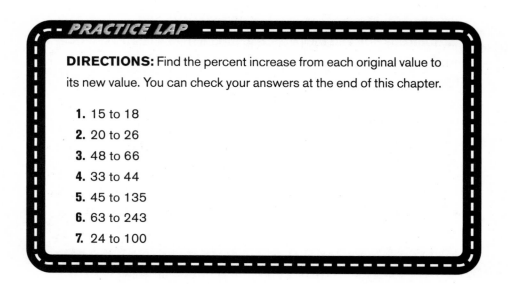

PRACTICE LAP

DIRECTIONS: Find the percent increase from each original value to its new value. You can check your answers at the end of this chapter.

1. 15 to 18
2. 20 to 26
3. 48 to 66
4. 33 to 44
5. 45 to 135
6. 63 to 243
7. 24 to 100

Finding the Value after a Percent Increase

An ice cream vendor sold 55 cones yesterday. If he wants to increase his sales by 40%, how many cones must he sell today? This is another way of looking at percent increase problems—sometimes you know the original value and the percent increase, and you need to find the new value.

Because **(percent increase) = (new value – original value) ÷ (original value),** we can rearrange this formula to solve for the new value:

(new value) = (percent increase × original value) + original value

Now let's plug the values from the ice cream vendor into this equation. The percent increase is 40%, or 0.40, and the original value is 55: (new value) = (0.40 × 55) + 55. 0.40 × 55 = 22, 22 + 55 = 77. If the vendor wants to increase his sales by 40%, he must sell 77 cones today.

CAUTION!

WHEN SUBSTITUTING A percent into a formula, be sure to convert it to a decimal before substituting it, or your answer could be wildly incorrect.

Example

If the value 40 increases by 32%, what is the new value?

Substitute the values into the formula:
$(0.32 \times 40) + 40 = 12.8 + 40 = 52.8$.

In these two examples, we used a formula derived from our percent increase formula to find the new values. However, there is an easier way: After converting the percent increase to a decimal, add 1 to it and multiply it by the original value. Let's use this method on the previous example.

The percent increase is 32%, or 0.32. Adding 1 gives us 1.32. $1.32 \times 40 = 52.8$.

Example

Trey has 60 comic books. After attending a convention, the size of his collection is now 15% larger. How many comic books does Trey have now?

The percent increase is 15%, or 0.15. Add 1 and multiply by the original value: $1.15 \times 60 = 69$. Trey has 69 comic books now.

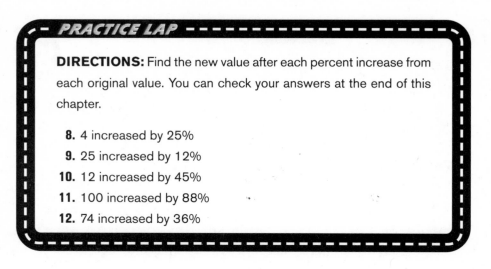

DIRECTIONS: Find the new value after each percent increase from each original value. You can check your answers at the end of this chapter.

8. 4 increased by 25%
9. 25 increased by 12%
10. 12 increased by 45%
11. 100 increased by 88%
12. 74 increased by 36%

And, as you might have guessed, if we have the new value and the percent increase, we can find the original value.

Example

DeDe read 8 magazines this week, 60% more magazines than she read last week. How many magazines did she read last week?

In this problem, 8 is the new value, 60% is the percent increase, and we are looking for the original value. We go back to our first formula: **(percent increase) = (new value – original value) ÷ (original value).** Because the original value appears twice in this formula, we'll let x hold the place of the original value, and we'll plug in 8 for the new value and 0.60 for the percent increase: $0.60 = (8 - x) ÷ x$.

Now we must solve for x, the original value. Multiplying both sides of the equation by x, we have $0.60x = 8 - x$. Add x to both sides of the equation, and it becomes $1.60x = 8$. Now divide both sides of the equation by 1.60, and x, the original value, is 5.

That's a lot of algebra, but fortunately, there is an easier way. We saw how we can find the new value, given the original value and the percent increase, by *multiplying* the original value by 1 plus the percent increase. We can find the original value, given the new value and the percent increase, by *dividing* the new value by 1 plus the percent increase.

Example

A value is increased by 5% to 42. What was the original value?

Write the percent increase as a decimal: 0.05. Add 1 to the percent increase and divide the new value by that sum: $42 \div 1.05 = 40$.

Example

A value is increased by 342% to 331.5. What was the original value?

Apply the same formula: $342\% = 3.42$, $3.42 + 1 = 4.42$. $256.5 \div 4.42 = 75$.

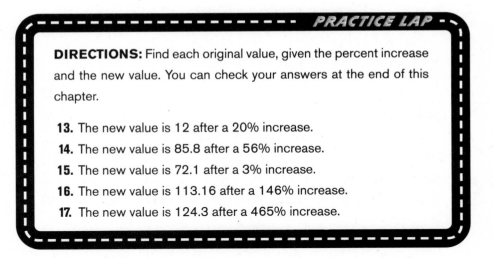

PRACTICE LAP

DIRECTIONS: Find each original value, given the percent increase and the new value. You can check your answers at the end of this chapter.

13. The new value is 12 after a 20% increase.

14. The new value is 85.8 after a 56% increase.

15. The new value is 72.1 after a 3% increase.

16. The new value is 113.16 after a 146% increase.

17. The new value is 124.3 after a 465% increase.

PERCENT DECREASE

Now we'll go in the other direction to find percent decrease. When a new value is *larger* than an original value, we have a percent increase, and when a new value is *smaller* than an original value, we have a **percent decrease**. Because of that fact, the formula we use for percent decrease is very similar to our percent increase formula, but with one change. Instead of subtracting the original value from the new value, we subtract the new value from the original value. We still divide by the original value.

FUEL FOR THOUGHT

PERCENT DECREASE, LIKE percent increase, is the difference between an original value and a new value divided by the original value. However, when finding a percent decrease, the new value is subtracted from the original value, because the original value is larger than the new value. Percent decrease is used to show the decline from an original value to a new value.

Example

Find the percent decrease from 18 to 15.

Subtract the new value from the original value and divide by the original value: $18 - 15 = 3$, $3 \div 18 = 0.1667$, to the nearest ten thousandth. $0.1667 = 16.67\%$.

CAUTION!

LOOK BACK AT the first practice question in this chapter. We found that the percent increase from 15 to 18 was 20%. However, we just found that the percent decrease from 18 to 15 was only 16.67%. How can that be? When we found the percent increase from 15 to 18, our original value was 15, so we divided the difference between the values by 15. When we found the percent decrease from 18 to 15, our original value was 18, so we divided the difference between the values by 18. As long as the two values are different (and if they are the same, then there is no percent increase or percent decrease), the percent increase from the first to the second will always be different than the percent decrease from the second to the first.

Let's look at one more example.

Example

Find the percent decrease from 29 to 22.

Subtract the new value from the original value and divide by the original value: 29 − 22 = 7, 7 ÷ 29 = 0.2414, to the nearest ten thousandth. 0.2414 = 24.14%.

It's also possible for a value to drop so far that it becomes negative.

Example

Find the percent decrease from 5 to −3.

We still use the same steps: subtract the new value, −3, from the original, and divide by the original. 5 − (−3) = 8, 8 ÷ 5 = 1.6 = 160%.

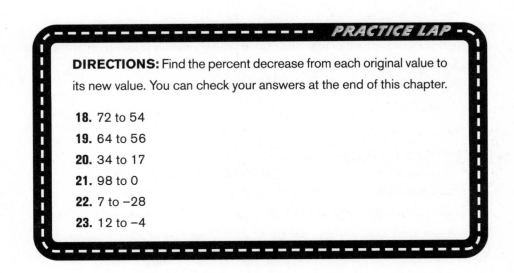

PRACTICE LAP

DIRECTIONS: Find the percent decrease from each original value to its new value. You can check your answers at the end of this chapter.

18. 72 to 54

19. 64 to 56

20. 34 to 17

21. 98 to 0

22. 7 to −28

23. 12 to −4

PACE YOURSELF

WRITE TEN NUMBERS on a sheet of paper. Draw lines connecting them, so that you have five pairs of numbers. Find the percent increase or decrease between the numbers in each pair. Then, look at all ten numbers. Between which two numbers would there be the largest percent increase? The largest percent decrease? Did you choose the same two numbers, but in reverse order?

Finding the Value after a Percent Decrease

To recap, we have two formulas for finding the new value, given an original value and a percent increase:

(new value) = (percent increase × original value) + original value
and
(new value) = (percent increase + 1) × (original value)

We also have two similar formulas for finding a new value, given an original value and a percent decrease:

(new value) = (original value) – (percent decrease × original value)
and
(new value) = (1 – percent decrease) × (original value)

Because the second formula is a bit simpler, we will use that.

Example

If the value 74 decreases by 35%, what is the new value?

The original value is 74 and the percent decrease is 35%, or 0.35. Substitute the values into the formula: $(1 - 0.35) \times 74 = 0.65 \times 74 = 48.1$.

Example

If the value 39 decreases by 120%, what is the new value?

Substitute the values into the formula:

$(1 - 1.20) \times 39 = -0.2 \times 39 = -7.8.$

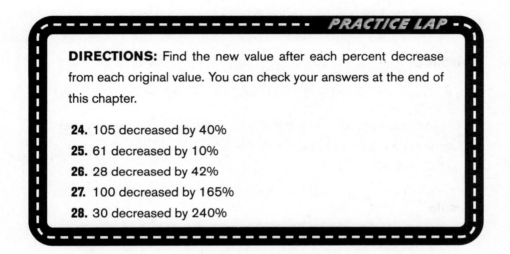

PRACTICE LAP

DIRECTIONS: Find the new value after each percent decrease from each original value. You can check your answers at the end of this chapter.

24. 105 decreased by 40%
25. 61 decreased by 10%
26. 28 decreased by 42%
27. 100 decreased by 165%
28. 30 decreased by 240%

All that's left is how to find the original value given a new value and a percent decrease.

Example

A glass of water has 6 ounces in it after 40% of the original volume has been drunk. What was the original volume of water in the glass?

In this problem, 6 is the new value, 40% is the percent decrease, and we are looking for the original value. We could go back to our original formula and substitute the two known values:

(percent decrease) = (original value − new value) ÷ (original value)

But we already know from finding the original value given a new value and a percent increase that this involves a lot of algebra. Just as there was an easier formula for that, there is an easier formula for this. We can find the original value, given the new value and the percent decrease, by dividing the new value by 1 minus the percent decrease:

(original value) = 6 ÷ (1 – 0.40) = 6 ÷ 0.6 = 10. The original volume of water in the glass was 10 ounces.

Example

A value is decreased by 33% to 14.07. What was the original value?

Write the percent decrease as a decimal: 0.33. Subtract the percent decrease from 1 and divide the new value by *that* difference: 14.07 ÷ 0.67 = 21.

Example

A value is decreased by 170% to –80.5. What was the original value?

Apply the same formula: 170% = 1.70, 1 – 1.70 = –0.70. –80.5 ÷ –0.70 = 115.

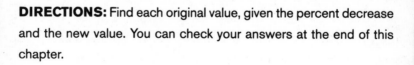

DIRECTIONS: Find each original value, given the percent decrease and the new value. You can check your answers at the end of this chapter.

29. The new value is 15 after a 70% decrease.

30. The new value is 23.1 after a 23% decrease.

31. The new value is 20.8 after an 80% decrease.

32. The new value is –50.16 after a 176% decrease.

33. The new value is –574 after a 510% decrease.

WHAT FORMULA TO USE?

Here are a few real-life percent increase and percent decrease problems. Let's look at each and decide which formula is needed to solve the problem.

Example

> A shirt at a clothing store normally costs $20, but is on sale for 10% off. What is the cost of the shirt?
>
> We are given the original value, $20, and the percent decrease. We are looking for the new value, so we use the formula **(new value) = (1 – percent decrease) × (original value):**
>
> $$\text{New value} = (1 - 0.10) \times \$20 = 0.9 \times \$20 = \$18.$$ The cost of the shirt is $18.

Example

> A sporting goods store increases the price of a baseball by 12% to $8.40. What was the original price of the baseball?
>
> We are given a new value, $8.40, and a percent increase. We are looking for the original value, so we use the formula **(original value) = (new value) ÷ (1 + percent increase).**
>
> Original value = $8.40 ÷ (1 + 0.12) = $8.40 ÷ 1.12 = $7.50. The original cost of the baseball was $7.50.
>
> Sometimes we need to use two formulas to solve a problem.

Example

> A television at Electronics World costs $149. The store increases the price by 8%, but when sales are slow, they reduce that price by 10%. What is the price of the television now?
>
> We need to use two formulas. First, we need to find the price of the television *after* an 8% increase: **(new value) = (percent**

increase + 1) × (**original value**), so the new value is equal to (.08 + 1) × $149 = 1.08 × $149 = $160.92. Now, we need to find the price of the television *after* a 10% decrease. (**new value**) = (1 – **percent decrease**) × (**original value**), so the new value is equal to (1 – 0.10) × $160.92 = 0.9 × $160.92 = $144.828, or $144.83.

Example

A value is decreased by 25%, and then increased by 7% to 54.57. What was the original value?

Work backward from the value 54.57. If the new value after a 7% increase is 54.57, we need to find the original value *before* that increase: (**original value**) = (**new value**) ÷ (**1 + percent increase**), so 54.57 ÷ 1.07 = 51. Continuing to work backward, 51 is now the new value. We need to find the original value *before* the 25% decrease. (**original value**) = (**new value**) ÷ (**1 – percent decrease**), so 51 ÷ (1 – 0.25) = 51 ÷ 0.75 = 68. The original value was 68.

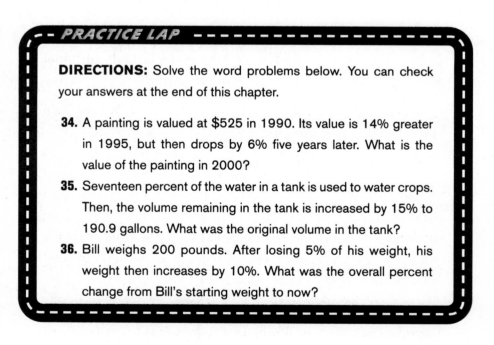

PRACTICE LAP

DIRECTIONS: Solve the word problems below. You can check your answers at the end of this chapter.

34. A painting is valued at $525 in 1990. Its value is 14% greater in 1995, but then drops by 6% five years later. What is the value of the painting in 2000?

35. Seventeen percent of the water in a tank is used to water crops. Then, the volume remaining in the tank is increased by 15% to 190.9 gallons. What was the original volume in the tank?

36. Bill weighs 200 pounds. After losing 5% of his weight, his weight then increases by 10%. What was the overall percent change from Bill's starting weight to now?

Applying Sales Tax

If a compact disc costs $15.99 and a shop charges 8% sales tax, what is the total cost of two compact discs?

When it comes to real-life percent problems, this is one of the most common: applying sales tax to a purchase. To solve a sales tax problem, begin by adding the costs of each item purchased. Then, convert the sales tax percent to a decimal, and add 1 to it. Finally, multiply that decimal by the cost of the items to find the total cost. Because we're working with money, round the total cost to the nearest hundredth (which is the nearest cent).

In the example above, two compact discs were purchased. The cost of the items is $15.99 × 2 = $31.98. The sales tax is 8%, or 0.08. If we add 1 to that, we will have 1 + 0.08 = 1.08. Now multiply that decimal by the cost of the items: $31.98 × 1.08 = $34.5384, or $34.54 after rounding to the nearest hundredth.

INSIDE TRACK

WHY DO WE add 1 to the sales tax decimal before multiplying? If we don't, we will find the total tax on the items, not the total cost of the items. We could add the cost of the items to the total tax, but multiplying by one more than the sales tax allows us to skip that step. However, if a problem asks for just the total tax, and not the total cost, then we must multiply by the sales tax decimal without adding 1.

Example

Nancy purchased a sandwich for $5.45 and a drink for $0.99. If she is charged 8.5% sales tax, what is the total cost of her purchase?

The total cost of the items is: $5.45 + $0.99 = $6.44. The sales tax, written as a decimal, is 0.085, and one greater than that is 1.085. Multiply the cost of the items by the sales tax: $6.44 × 1.085 = $6.9874, which is $6.99 after rounding.

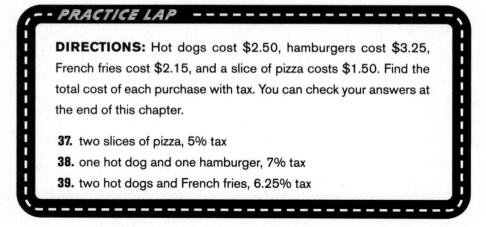

PRACTICE LAP

DIRECTIONS: Hot dogs cost $2.50, hamburgers cost $3.25, French fries cost $2.15, and a slice of pizza costs $1.50. Find the total cost of each purchase with tax. You can check your answers at the end of this chapter.

37. two slices of pizza, 5% tax

38. one hot dog and one hamburger, 7% tax

39. two hot dogs and French fries, 6.25% tax

Leaving a Tip

Now let's go one step further: After finding the total cost of a restaurant bill, determine how much to tip. It's often customary to tip 15% of the total bill after tax. Just as in the previous examples, begin by adding the cost of the items purchased. Then, convert the tax to a decimal and add 1 to it. Multiply that decimal by the cost of the items. This will give you the total bill. Finally, multiply that amount by 15% (0.15). This will tell you how much to tip.

Example

Ariscielle orders pancakes, which cost $4.75, and a cup of coffee, which costs $1.10. If she is charged 8% sales tax, what should she tip her waiter?

Add the cost of the items: $4.75 + $1.10 = $5.85. Write the sales tax as a percent and add 1 to it: 8% = 0.08, 1 + 0.08 = 1.08. Multiply that decimal by the cost of the items: $5.85 × 1.08 = $6.318, or $6.32. To find the tip, multiply the bill by 15%: $6.32 × 0.15 = $0.948, or $0.95. Often, the tip is rounded to the nearest dollar instead of to the nearest cent. Ariscielle should tip $1.00.

CAUTION!

ALTHOUGH WE ADD 1 to the sales tax decimal before multiplying to find the total bill, do not add 1 to the tip percent. When finding the tip amount, you are looking for 15% of the bill, not the total cost of the meal including tip. If you wanted to know what a meal cost you in total (items, tax, and tip), then you would multiply the bill after tax by 1.15.

Example

For dinner, Heather eats lasagna, which costs $7.75, and Serge has steak, which costs $13.99. They each drink a glass of milk, which costs $1.35 per glass. If the tax on the meal is 7.5%, to the nearest dollar, how much should they tip?

Add the cost of the lasagna and the cost of the steak to twice the cost of a glass of milk: $7.75 + $13.99 + 2 × $1.35 = $24.44. Now apply the sales tax. 7.5% = 0.075, 1 + 0.075 = 1.075. The total bill is equal to: $24.44 1.075 = $26.273, or $26.27. The tip should be 15% of that amount. 15% = 0.15. Multiply the total bill by the tip percent: $26.27 × 0.15 = $3.9405, or $4.00, rounded to the nearest dollar. Heather and Serge should tip $4.00.

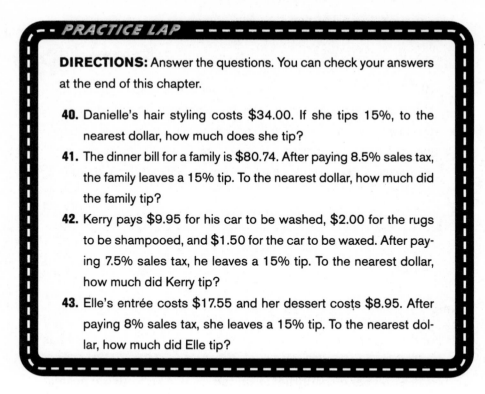

- - PRACTICE LAP

DIRECTIONS: Answer the questions. You can check your answers at the end of this chapter.

40. Danielle's hair styling costs $34.00. If she tips 15%, to the nearest dollar, how much does she tip?

41. The dinner bill for a family is $80.74. After paying 8.5% sales tax, the family leaves a 15% tip. To the nearest dollar, how much did the family tip?

42. Kerry pays $9.95 for his car to be washed, $2.00 for the rugs to be shampooed, and $1.50 for the car to be waxed. After paying 7.5% sales tax, he leaves a 15% tip. To the nearest dollar, how much did Kerry tip?

43. Elle's entrée costs $17.55 and her dessert costs $8.95. After paying 8% sales tax, she leaves a 15% tip. To the nearest dollar, how much did Elle tip?

Working with Simple Interest

Banks pay interest on the money a person deposits in an account. That deposit is called the **principal**, and the **interest** that the bank pays is based on the **rate** that the bank sets and the amount of time the principal is in the bank. The formula for finding interest is $I = prt$, where I is interest, p is principal, r is the rate, and t is the time in years. Given any three of those values, we can find the fourth.

To find the interest given the principal, rate, and time, convert the rate from a percent to a decimal, and then multiply the principal by the rate by the time.

Example

What is the interest on a principal of $500 at a rate of 5% for one year?

Convert the rate from a percent to a decimal. 5% = 0.05. Multiply the principal by the rate by the time: $500 × 0.05 × 1 = $25.

The formula $I = prt$ can be rewritten to find the principal given the interest, rate, and time: $p = I \div rt$.

Example

Robert gains $13.20 in interest after keeping his principal in a bank for 2 years at a rate of 3%. What was his principal?

Use the formula $p = I \div rt$. Divide the interest by the product of the rate and the time: $13.20 \div (0.03 \times 2) = 13.20 \div 0.06 = 220$. Robert's principal was $220.

The formula $I = prt$ can also be rewritten to find the time given the interest, rate, and principal: $t = I \div pr$.

Example

A principal of $390 at a rate of 3.5% gains $68.25 in interest. For how long was the principal in the bank?

Use the formula $t = I \div pr$. Divide the interest by the product of the principal and the rate: $68.25 \div (390 \times 0.035) = 68.25 \div 13.65 = 5$. The principal was in the bank for 5 years.

And finally, we can rewrite the interest formula to find the rate given the interest, principal, and time: $r = I \div pt$.

Example

A principal of $250 gains $42 in interest over 4 years. At what rate did the principal gain interest?

Use the formula $r = I \div pt$. Divide the interest by the product of the principal and the time: $42 \div (250 \times 4) = 42 \div 1{,}000 = 0.042 = 4.2\%$. The principal gained interest at a rate of 4.2%.

CAUTION!

THE FORMULA *I = prt* (and all the formulas that are related to it) assumes that the time is given in years. If you are given a time in months, you must convert it to years before using the formula.

Example

What is the interest on a principal of $630 at a rate of 5.5% for 6 months?

First, convert the time from months to years. Because there are 12 months in a year, 6 months is 6 ÷ 12 = 0.5 years. Next, convert the rate from a percent to a decimal: 5.5% = 0.055. Now we are ready to use the formula. Multiply the principal by the rate by the time: $630 × 0.055 × 0.5 = $17.325, or $17.33.

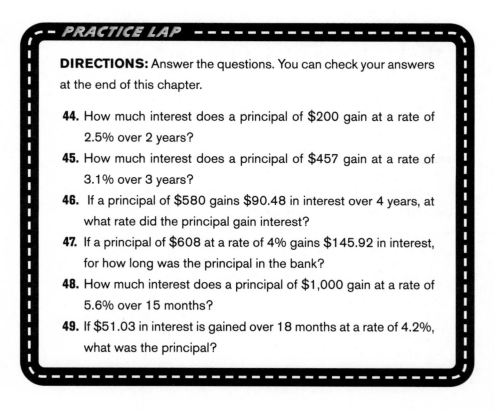

PRACTICE LAP

DIRECTIONS: Answer the questions. You can check your answers at the end of this chapter.

44. How much interest does a principal of $200 gain at a rate of 2.5% over 2 years?

45. How much interest does a principal of $457 gain at a rate of 3.1% over 3 years?

46. If a principal of $580 gains $90.48 in interest over 4 years, at what rate did the principal gain interest?

47. If a principal of $608 at a rate of 4% gains $145.92 in interest, for how long was the principal in the bank?

48. How much interest does a principal of $1,000 gain at a rate of 5.6% over 15 months?

49. If $51.03 in interest is gained over 18 months at a rate of 4.2%, what was the principal?

We can now use percents to handle everyday situations, such as leaving a tip or finding simple interest. You've seen how to work with fractions, decimals, and percents. In the next chapter, we'll put all three together, converting and comparing among them.

ANSWERS

1. $18 - 15 = 3, 3 \div 15 = 0.2 = 20\%$

2. $26 - 20 = 6, 6 \div 20 = 0.3 = 30\%$

3. $66 - 48 = 18, 18 \div 48 = 0.375 = 37.5\%$

4. $44 - 33 = 11, 11 \div 33 = 0.3333$ to the nearest ten thousandth: $0.3333 = 33.33\%$

5. $135 - 45 = 90, 90 \div 45 = 2 = 200\%$

6. $243 - 63 = 180, 180 \div 63 = 2.8571$ to the nearest ten thousandth: $2.8571 = 285.71\%$

7. $100 - 24 = 76, 76 \div 24 = 3.1667$ to the nearest ten thousandth: $3.1667 = 316.67\%$

8. $25\% = 0.25, 4 \times 1.25 = 5$

9. $12\% = 0.12, 25 \times 1.12 = 28$

10. $45\% = 0.45, 12 \times 1.45 = 17.4$

11. $88\% = 0.88, 100 \div 1.88 = 188$

12. $36\% = 0.36, 74 \div 1.36 = 100.64$

13. $20\% = 0.2, 12 \div 1.2 = 10$

14. $56\% = 0.56, 85.8 \div 1.56 = 55$

15. $3\% = 0.03, 72.1 \div 1.03 = 70$

16. $146\% = 1.46, 113.16 \div 2.46 = 46$

17. $465\% = 4.65, 124.3 \div 5.65 = 22$

18. $72 - 54 = 18, 18 \div 72 = 0.25 = 25\%$

19. $64 - 56 = 8, 8 \div 64 = 0.125 = 12.5\%$

20. $34 - 17 = 17, 17 \div 34 = 0.5 = 50\%$

21. $98 - 0 = 98, 98 \div 98 = 1 = 100\%$

22. $7 - (-28) = 35, 35 \div 7 = 5 = 500\%$

23. $12 - (-4) = 16$, $16 \div 12 = 1.3333$ to the nearest ten thousandth. $1.3333 = 133.33\%$.

24. $(1 - 0.40) \times 105 = 0.6 \times 105 = 63$

25. $(1 - 0.10) \times 61 = 0.9 \times 61 = 54.9$

26. $(1 - 0.42) \times 28 = 0.58 \times 28 = 16.24$

27. $(1 - 1.65) \times 100 = -0.65 \times 100 = -65$

28. $(1 - 2.40) \times 30 = -1.4 \times 30 = -42$

29. $70\% = 0.7$, $1 - 0.7 = 0.3$. $\times 15 \div 0.3 = 50$

30. $23\% = 0.23$, $1 - 0.23 = 0.77$. $\times 23.1 \div 0.77 = 30$

31. $80\% = 0.8$, $1 - 0.8 = 0.2$. $\times 20.8 \div 0.2 = 104$

32. $176\% = 1.76$, $1 - 1.76 = -0.76$. $\times -50.16 \div -0.76 = 66$

33. $510\% = 5.1$, $1 - 5.1 = -4.1$. $\times -574 \div -4.1 = 140$

34. First, the value of the painting increases by 14%: **(new value)** = **(original value)** $\times$ **(1 + percent increase)**, so $\$525 \times 1.14 = \$598.50$. Then, the value of the painting decreases by 6%: **(new value)** = **(original value)** $\times$ **(1 – percent decrease)**, so $\$598.50 \times 0.94 = \$562.59$.

35. Work backward: First, find the volume in the tank *before* it was increased by 15% to 190.9 gallons: **(original value)** = **(new value)** $\div$ **(1 + percent increase)**, so $190.9 \div 1.15 = 166$ gallons. Now, find the original volume in the tank *before* it was decreased by 17%: **(original value)** = **(new value)** $\div$ **(1 – percent decrease)**, so $166 \div 0.83 = 200$ gallons.

36. Find Bill's weight after each change, and then compare his end weight to his starting weight to find the percent change. First, Bill's weight decreases by 5%: **(new value)** = **(original value)** $\times$ **(1 – percent decrease)**, so $200 \times 0.95 = 190$ pounds. Then, Bill's weight increases by 10%: **(new value)** = **(original value) (1 + percent increase)**, so $190 \times 1.10 = 209$ pounds. Bill's original weight was 200 pounds and his new weight is 209 pounds: **(percent increase)** = **(new value – original value)** $\div$ **(original value)**, so $(209 - 200) \div 200 = 9 \div 200 = 0.045$, or 4.5%. Bill's weight increased by 4.5%.

37. One slice of pizza costs $\$1.50$. $\$1.50 \times 2 = \$3.00$. $5\% = 0.05$, $1 + 0.05 = 1.05$. $\$3.00 \times 1.05 = \$3.15$.

38. One hot dog costs $2.50 and one hamburger costs $3.25. $2.50 + $3.25 = $5.75. 7% = 0.07, 1 + 0.07 = 1.07. $5.75 × 1.07 = $6.1525, or $6.15.

39. One hot dog costs $2.50 and French fries cost $2.15. $2.50 × 2 = $5.00, $5.00 + $2.15 = $7.15. 6.25% = 0.0625, 1 + 0.0625 = 1.0625. $7.15 × 1.0625 = $7.596875, or $7.60.

40. $34.00 × 0.15 = $5.10, or $5 to the nearest dollar.

41. $80.74 × 1.085 = $87.6029, or $87.60 to the nearest cent. $87.60 × 0.15 = $13.14, or $13 to the nearest dollar.

42. $9.95 + $2.00 + $1.50 = $13.45. $13.45 × 1.075 = $14.45875, or $14.46 to the nearest cent. $14.46 × 0.15 = $2.169, or $2 to the nearest dollar.

43. $17.55 + $8.95 = $26.50. $26.50 × 1.08 = $28.62. $28.62 × 0.15 = $4.293, or $4 to the nearest dollar.

44. Use the formula $I = prt$: $200 × 0.025 × 2 = $10.

45. Use the formula $I = prt$: $457 × 0.031 × 3 = $42.501, or $42.50.

46. Use the formula $r = I \div pt$: $90.48 ($580 × 4) = $90.48 ÷ 2,320 = 0.039, or 3.9%.

47. Use the formula $t = I \div pr$: $145.92 ($608 × 0.04) = $145.92 ÷ $24.32 = 6 years.

48. The time is given in months, so it must be converted to years: 15 ÷ 12 = 1.25. Now we can use the formula $I = prt$: $1,000 × 0.056 × 1.25 = $70.

49. The time is given in months, so it must be converted to years. 18 × 12 = 1.5. Now we can use the formula $p = I \div rt$: $51.03 ÷ (0.042 × 1.5) = $51.03 ÷ 0.063 = $810.

From Fractions, to Decimals, to Percents, and Back

WHAT'S AROUND THE BEND

- ➡ Writing Fractions and Mixed Numbers as Decimals
- ➡ Writing Decimals as Percents
- ➡ Writing Fractions and Mixed Numbers as Percents
- ➡ Writing Decimals as Fractions and Mixed Numbers
- ➡ Writing Percents as Decimals
- ➡ Writing Percents as Fractions and Mixed Numbers
- ➡ Comparing Fractions, Decimals, and Percents
- ➡ Ordering Fractions, Decimals, and Percents

In the last two chapters, we learned how to write percents as decimals in order to work with them. Not only can percents be turned into decimals, but any number that can be written as a percent can also be written as a fraction. In this chapter, we'll learn how to write fractions as decimals and percents, how to write decimals as fractions and percents, and how to write percents as fractions and decimals.

WRITING FRACTIONS AND MIXED NUMBERS AS DECIMALS

By now, it probably seems like you've read this ten times: Fractions represent division. The fraction $\frac{1}{2}$ means "1 divided by .2." To convert a fraction to a decimal, divide the numerator by the denominator. That's all there is to it.

Example

Write $\frac{3}{4}$ as a decimal.

Divide the numerator, 3, by the denominator, 4: $3 \div 4 = 0.75$. $\frac{3}{4} = 0.75$.

A funny thing happens when we write a fraction that has a denominator of 10 or 100 (or any power of 10) as a fraction.

Example

Write $\frac{7}{10}$ as a decimal.

Divide the numerator, 7, by the denominator, 10: $7 \div 10 = 0.7$. $\frac{7}{10} = 0.7$. The numerator of the fraction is the tenths digit of the decimal. This isn't a big surprise; both $\frac{7}{10}$ and 0.7 are read as "seven tenths." As we learned earlier, each decimal place is a power of ten. Any fraction with a denominator that is a power of ten can be written as a decimal without dividing.

Example

Write $\frac{4}{1,000}$ as a decimal.

Write the number 4 in the thousandths place: 0.004. If the numerator of the fraction has more than one digit, use the places to the left of the thousandths place. $\frac{56}{1,000} = 0.056$ and $\frac{237}{1,000}$ is equal to 0.237.

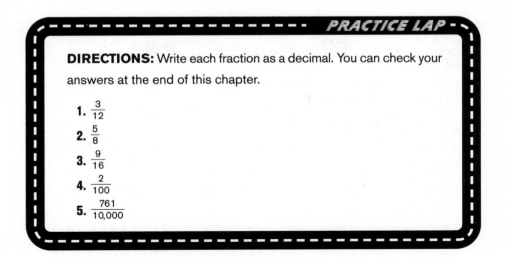

PRACTICE LAP

DIRECTIONS: Write each fraction as a decimal. You can check your answers at the end of this chapter.

1. $\frac{3}{12}$

2. $\frac{5}{8}$

3. $\frac{9}{16}$

4. $\frac{2}{100}$

5. $\frac{761}{10,000}$

Although every fraction can be written as a decimal, not every fraction can be written as a terminating decimal. The decimals we have looked at so far in this chapter have all been **terminating decimals**, because we can divide the numerator by the denominator and complete our division.

FUEL FOR THOUGHT

A TERMINATING DECIMAL is a decimal that does not continue infinitely. In other words, a terminating decimal is a value that can be represented fully with a finite number of digits. For instance, 0.25 is a terminating decimal.

When we try to express some fractions, such as $\frac{1}{3}$, as decimals, we find that the decimal continues on and on and we are never able to complete our division. These types of decimals are called **non-terminating decimals**.

FUEL FOR THOUGHT

A NON-TERMINATING DECIMAL is a decimal whose digits continue forever. A non-terminating decimal may repeat, such as 0.3333. . . . If a non-terminating decimal does not repeat, then it is an irrational number. All rational numbers can be written as either terminating decimals or repeating decimals. A **rational number** can be written in the form $\frac{x}{y}$, where x and y are integers and y is not equal to zero. An **irrational number** is a number that cannot be written in that form.

Fractions that are non-terminating decimals are a specific kind of non-terminating decimals, called **repeating decimals**. The fraction $\frac{1}{3}$ is a repeating decimal because the number three repeats over and over, indefinitely. $\frac{1}{3} =$ 0.3333. . . Sometimes, to show that a digit or group of digits repeats, we put a bar over the repeating digits: $\frac{1}{3} = 0.$

FUEL FOR THOUGHT

A REPEATING DECIMAL is a decimal in which the same digit or sequence of digits repeats over and over, infinitely. A single digit could repeat, such as 0.33333. . ., or a sequence of digits might repeat, such as 0.297297297. . . . The repetition can begin before or after the decimal point, and not necessarily at the first digit after the decimal point (the tenths digit). For example, the decimal 0.916666666. . . begins to repeat starting with the thousandths digit. A repeating decimal is an example of a non-terminating decimal.

All fractions can be written as either terminating decimals or repeating decimals.

Example

Write $\frac{1}{27}$ as a decimal.

If we divide 1 by 27, we get the same three digits over and over: 0.037037037. . . This can be written as 0.037, because the digits 037 repeat infinitely. We could also round the decimal to the thousandths place, and simply put 0.037.

CAUTION!

IN THE EXAMPLES $\frac{1}{3}$ and $\frac{1}{27}$, all of the digits to the right of the decimal point repeated. That is not always true—sometimes, the digits may not begin to repeat until the hundredths place, thousandths place, or even farther to the right of the decimal point. For example, the fraction $\frac{1}{6}$ is equal to 0.1666666666. . . The digit 6 continues to repeat. In this decimal, the repeating part begins in the hundredths place.

The repeating digits in some decimals are so long, it is often better to round than to try to show all the digits that repeat.

Example

Write $\frac{21}{201}$ as a decimal.

If we divide 21 by 201, the first eight digits to the right of the decimal point are 0.10447761. We know this decimal must repeat or terminate because all fractions can be expressed as terminating or repeating decimals, but it may be many, many digits before this decimal begins to repeat. For these types of decimals, it may be best to round. $\frac{21}{201}$, to the nearest thousandth, is 0.104.

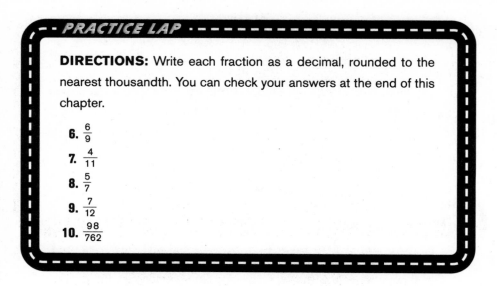

DIRECTIONS: Write each fraction as a decimal, rounded to the nearest thousandth. You can check your answers at the end of this chapter.

6. $\frac{6}{9}$

7. $\frac{4}{11}$

8. $\frac{5}{7}$

9. $\frac{7}{12}$

10. $\frac{98}{762}$

Writing mixed numbers as decimals is very similar. The whole number part of a mixed number is written to the left of the decimal point. And the fractional part of a mixed number? We just learned how to write that: by dividing the numerator by the denominator.

Example

Write $4\frac{1}{4}$ as a decimal.

Write 4 to the left of the decimal point and divide 1 by 4: $1 \div 4 = 0.25$, so $4\frac{1}{4} = 4.25$.

Example

Write $26\frac{6}{13}$ as a decimal.

Write 26 to the left of the decimal point and divide 6 by 13. $6 \div 13 = 0.461538. . .$, so $26\frac{6}{13} = 26.462$, to the nearest thousandth.

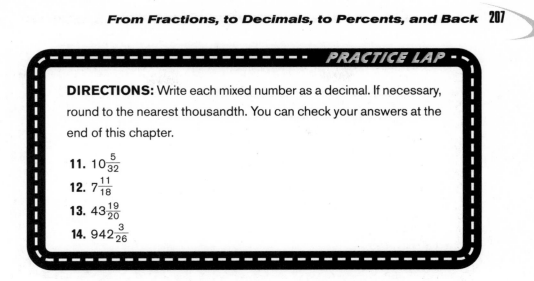

PRACTICE LAP

DIRECTIONS: Write each mixed number as a decimal. If necessary, round to the nearest thousandth. You can check your answers at the end of this chapter.

11. $10\frac{5}{32}$

12. $7\frac{11}{18}$

13. $43\frac{19}{20}$

14. $942\frac{3}{26}$

QUICK REVIEW: WRITING DECIMALS AS PERCENTS

In Chapter 9, we learned that we could write a decimal as a percent by moving the decimal point of a number two places to the right and adding the percent sign (%).

Example

Write 0.65 as a percent.

Move the decimal point two places to the right and add the percent sign: 0.65 = 65%.

Example

Write 0.045 as a percent.

Percents can also have a decimal point. After moving the decimal point two places to the right, we find that 0.045 = 4.5%.

Example

Write 5.6103 as a percent.

Percents can be 100 or greater. After moving the decimal point two places to the right, we find that 5.6103 = 561.03%.

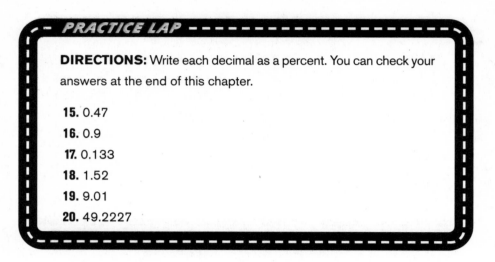

PRACTICE LAP

DIRECTIONS: Write each decimal as a percent. You can check your answers at the end of this chapter.

15. 0.47

16. 0.9

17. 0.133

18. 1.52

19. 9.01

20. 49.2227

WRITING FRACTIONS AND MIXED NUMBERS AS PERCENTS

Now that we know how to write fractions and mixed numbers as decimals, and we know how to write decimals as percents, we can write fractions and mixed numbers as percents by putting the two steps together. Write a fraction as a decimal, and then turn that decimal into a percent.

Example

Write $\frac{2}{5}$ as a percent.

First, write the fraction as a decimal by dividing: $2 \div 5 = 0.4$. Now, write the decimal as a percent by moving the decimal point two places to the right: $0.4 = 40\%$, which means that $\frac{2}{5} = 40\%$.

What about fractions that are repeating decimals? We rounded those decimals to the nearest thousandth, and we can convert that rounded value to a percent.

Example

Write $\frac{11}{15}$ as a percent.

Write the fraction as a decimal by dividing. $11 \div 15 = 0.733333\ldots$, or 0.733 to the nearest thousandth. Now, move the decimal point two places to the right. 0.733 = 73.3%. The fraction $\frac{11}{15}$ is equal to approximately 73.3%.

Mixed numbers can also be written as percents. Be sure to write the whole number part of the decimal before moving the decimal point.

Example

Write $14\frac{6}{25}$ as a percent.

Write the fractional part as a decimal by dividing: $6 \div 25 = 0.24$, so $14\frac{6}{25} = 14.24$. Move the decimal point two places to the right: $14.24 = 1{,}424\%$. The mixed number $14\frac{6}{25}$ is equal to 1,424%.

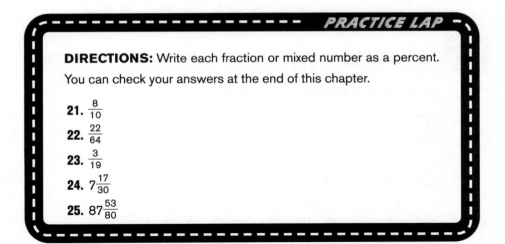

PRACTICE LAP

DIRECTIONS: Write each fraction or mixed number as a percent. You can check your answers at the end of this chapter.

21. $\frac{8}{10}$

22. $\frac{22}{64}$

23. $\frac{3}{19}$

24. $7\frac{17}{30}$

25. $87\frac{53}{80}$

WRITING DECIMALS AS FRACTIONS AND MIXED NUMBERS

Now, let's go in the other direction: We'll start with decimals and turn them into fractions. When given a decimal number, read the decimal number out loud. The name of a decimal number is the same as its fraction name. The first part of the name will be the numerator of the fraction and the second part of the name will be the denominator of the fraction.

Example

Write 0.3 as a fraction.

Read the name of the decimal out loud. 0.3 is "three tenths." The first part of the name, *three,* is the numerator of the fraction. The second part of the name, *tenths,* is the denominator of the fraction: $0.3 = \frac{3}{10}$.

Example

Write 0.81 as a fraction.

Again, read the name of the decimal out loud. 0.81 is "eighty-one hundredths." The numerator of the fraction is 81. The denominator of the fraction is 100: $0.81 = \frac{81}{100}$.

CAUTION!

BE CAREFUL WHEN converting a decimal to a fraction. Sometimes, a decimal might "look" like a common fraction. For example, 0.333 "looks" like $\frac{1}{3}$. However, 0.333 is a terminating decimal that has an exact value of $\frac{333}{1,000}$. $\frac{1}{3}$ is equal to 0.333. . . or $0.\overline{3}$, which is a repeating decimal. 0.333 is *not* equal to $\frac{1}{3}$. Always read the decimal out loud, and then write its exact numerator and denominator.

The same strategy works for decimals that are greater than or equal to 1.

Example

Write 6.795 as a mixed number.

6.795 is "six and seven hundred ninety-five thousandths." The number to the left of the decimal point, which is read before the word *and* in the name of the decimal, is the whole number part of the mixed number. The numerator of the fraction is 795. The denominator of the fraction is 1,000: $6.795 = 6\frac{795}{1,000}$, which simplifies to $6\frac{159}{200}$.

INSIDE TRACK

THE NUMERATOR OF the fraction will always be exactly what you see to the right of the decimal point. We read the decimal out loud to help us figure out what the denominator of the fraction will be. The denominator of the fraction will come from the name of the rightmost decimal place in the number. For example, the decimal 0.78 will have a numerator of 78. The rightmost decimal place is the hundredths place, so the denominator of the fraction will be 100.

After writing a decimal as a fraction, we can often simplify the fraction. If the numerator is divisible by 2 or 5, then the fraction can be simplified because every power of 10 is divisible by 2 and 5.

Example

Write 0.25 as a fraction in simplest form.

0.25 is "twenty-five hundredths." The numerator of the fraction is 25. The denominator of the fraction is 100: $0.25 = \frac{25}{100}$. The greatest common factor of 25 and 100 is 25. Divide the numerator and denominator of the fraction by 25: $\frac{25}{100} = \frac{1}{4}$.

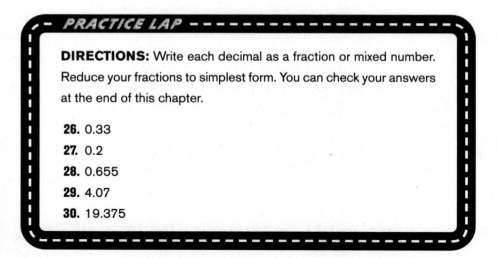

PRACTICE LAP

DIRECTIONS: Write each decimal as a fraction or mixed number. Reduce your fractions to simplest form. You can check your answers at the end of this chapter.

26. 0.33
27. 0.2
28. 0.655
29. 4.07
30. 19.375

QUICK REVIEW: WRITING PERCENTS AS DECIMALS

In Chapter 9, we learned that we could write a percent as a decimal by moving the decimal point of a number two places to the left and removing the percent sign (%).

Example

Write 14% as a decimal.

Move the decimal point two places to the left and remove the percent sign: 14% = 0.14.

Example

Write 2.654% as a decimal.

Move the decimal point two places to the left and remove the percent sign: 2.654% = 0.02654.

We can also write percents that are greater than 100 or less than 1 as decimals. No matter where the decimal point is in the percent, move it two places to the left to turn it into a decimal.

Example

Write 148.8% as a decimal.

148.8% = 1.488.

Example

Write 0.166% as a decimal.

0.166% = 0.00166.

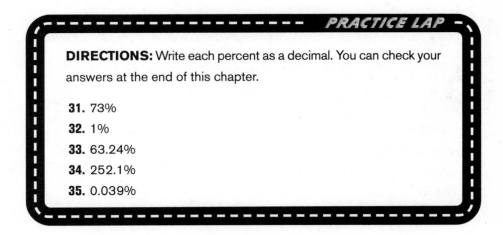

PRACTICE LAP

DIRECTIONS: Write each percent as a decimal. You can check your answers at the end of this chapter.

31. 73%

32. 1%

33. 63.24%

34. 252.1%

35. 0.039%

WRITING PERCENTS AS FRACTIONS AND MIXED NUMBERS

Just as we converted fractions and mixed numbers to percents by first converting them to decimals, and then converting the decimals to percents, we go in the opposite direction to convert percents to fractions and mixed numbers. Begin by writing the percent as a decimal, and then write the decimal as a fraction or mixed number.

Example

Write 31% as a fraction.

First, write 31% as a decimal. Move the decimal point two places to the left and remove the percent sign: 31% = 0.31. Now, write 0.31 as a fraction. 0.31 is "thirty-one hundredths," which means that the numerator of the fraction is 31 and the denominator of the fraction is 100: $31\% = 0.31 = \frac{31}{100}$.

Example

Write 97.201% as a fraction.

Write 97.201% as a decimal: $97.201\% = 0.97201$. Now, write 0.97201 as a fraction. 0.97201 is "ninety-seven thousand, two hundred one hundred thousandths," which means that the numerator of the fraction is 97,201 and the denominator of the fraction is 100,000: $97.201\% = 0.97201 = \frac{97,201}{100,000}$.

CAUTION!

DO NOT TRY to convert a fraction directly to a percent, or convert a percent directly to a fraction. We never truly convert fractions to percents or percents to fractions. In both cases, we convert to decimals first.

PRACTICE LAP

DIRECTIONS: Write each percent as a fraction or mixed number. Reduce your fractions to simplest form. You can check your answers at the end of this chapter.

36. 9%

37. 26%

38. 88.1%

39. 429%

40. 0.667%

PACE YOURSELF

ANY NUMBER CAN be converted from a decimal to a percent or fraction. Roll a number cube three times and mark down the number you rolled each time. Make the result of your first roll the tenths digit of a number, make the result of the second roll the hundredths digit of the number, and make the result of the third roll the thousandths digit of the number. Write that number as a fraction and as a percent.

COMPARING FRACTIONS, DECIMALS, AND PERCENTS

We learned how to compare fractions to fractions in Chapter 3, and we learned how to compare decimals to decimals in Chapter 7. Now, we'll compare fractions to decimals, decimals to percents, and percents to fractions. How? By first converting any fractions or percents to decimals. It is easiest to compare a decimal to another decimal.

Example

Compare $\frac{1}{8}$ and 0.2.

Convert the fraction $\frac{1}{8}$ to a decimal. $1 \div 8 = 0.125$. Now, compare 0.125 to 0.2 using the strategy we learned in Chapter 7. First, line up the decimal points and add two trailing zeros to 0.2:

$$0.125$$
$$0.200$$

Next, compare digits from left to right. Both numbers have a zero in the ones place. The first number, 0.125, has a 1 in the tenths place, and the second number, 0.200, has a 2 in the tenths place. Because $2 > 1$, 0.2 is greater than 0.125, and 0.2 is greater than $\frac{1}{8}$.

Example

Compare 45% and 2.3.

45% looks larger at first glance, but once we convert it to a decimal by moving the decimal point two places to the left, we see that 45% = 0.45. Line up the decimal points, add a trailing zero to 2.3, and compare:

0.45

2.30

2.3 has a 2 in the ones place and 0.45 has a 0 in the ones place, so 2.3 is greater than 45%.

CAUTION!

BECAUSE 45% IS less than 100%, you may have realized that it was less than 1, and therefore less than 2.3, without even converting it to a decimal. However, the safest strategy is still to convert numbers to decimals before comparing. By comparing decimals digit by digit from left to right, the process is reduced to simple steps, and you are less likely to make a mistake.

Even if neither number being compared is a decimal, we still convert both numbers to decimals.

Example

Compare 17% and $\frac{1}{7}$.

17% = 0.17 and $\frac{1}{7}$ = 1 ÷ 7 = 0.142857. . ., or 0.143 to the nearest thousandth. Line up the decimal points, add a trailing zero to 0.17, and compare:

0.170

0.143

Both numbers have the same digits in the ones and tenths places, but 0.170 has a 7 in the hundredths place and 0.143 has a 4 in the hundredths place. $7 > 4$, and 0.17 is greater than $\frac{1}{7}$.

Example

Compare 8.54 and $8\frac{1}{2}$.

$8\frac{1}{2} = 8.5$. Line up the decimal points, add a trailing zero to 8.5, and compare:

$$8.54$$
$$8.50$$

Both numbers have an 8 in the ones place and a 5 in the tens place. The first number, 8.54, has a 4 in the hundredths place, and the second number, 8.5, has a 0 in the hundredths place, so 8.54 is greater than 8.5, or $8\frac{1}{2}$.

Those numbers were almost equal. It wasn't until we reached the last (rightmost) decimal place that we were able to determine which number was greater. Sometimes, when comparing fractions, decimals, and percents, we will find two values that are equal.

Example

Compare 82.5% and $\frac{33}{40}$.

Convert each number to a decimal. $82.5\% = 0.825$. $\frac{33}{40} = 33 \div 40 = 0.825$. These two numbers are equal to each other.

Example

Compare 0.22 and $\frac{2}{9}$.

$\frac{2}{9} = 2 \div 9 = 0.222...$, or 0.222 to the nearest thousandth. Line up the decimal points and compare:

$$0.220$$
$$0.222$$

0.222 is greater than 0.220 because its thousandths digit is greater.

CAUTION!

WE HAVE OFTEN rounded repeating decimals to the thousandths place. However, if the decimal to which a repeating decimal is being compared has a thousandths digit, or digits farther to the right of the decimal point, we cannot round to the thousandths place, because it may lead to an incorrect comparison. When comparing a terminating decimal to a repeating decimal, always round the repeating decimal to the place that is one place farther to the right than the last digit of the terminating decimal.

Example

Compare $\frac{9}{14}$ and 0.64286.

$\frac{9}{14} = 9 \div 14 = 0.64285714...$, or 0.643 to the nearest thousandth. However, the decimal 0.64286 has a digit in the thousandths place, and digits farther to the right of the thousandths place. In fact, it has a digit in the hundred thousandths place. Therefore, we must round $\frac{9}{14}$ to the place to the right of the hundred thousandths place: We must round to the millionths place.

0.64285714 to the nearest millionth is 0.642857. Now we are ready to compare:

> 0.642857
>
> 0.642860

    The two numbers have the same digits until we reach the hundred thousandths place. The decimal 0.64286 has a larger hundred thousandths digit, so 0.64286 is greater than $\frac{9}{14}$. Notice that if we had rounded $\frac{9}{14}$ to the nearest thousandth, we would have incorrectly believed that it was greater than 0.64286.

PRACTICE LAP

DIRECTIONS: Place $>$, $<$ or $=$ between each pair of numbers. You can check your answers at the end of this chapter.

41. 0.596 ___ 5.96%

42. $\frac{17}{20}$ ___ 0.59

43. 72% ___ $\frac{3}{4}$

44. $\frac{21}{70}$ ___ 30%

45. 4.29 ___ $4\frac{2}{9}$

46. 0.5238 ___ $\frac{11}{21}$

47. 0.001 ___ 0.1%

48. $7\frac{5}{7}$ ___ 7.143

49. 29.45% ___ 2,945

50. $\frac{2}{3}$ ___ 67%

ORDERING FRACTIONS, DECIMALS, AND PERCENTS

We order fractions, decimals, and percents just as we ordered decimals in Chapter 7, because we convert fractions and percents to decimals before comparing and ordering them. Given a set of fractions, decimals, or percents, convert any fractions or percents to decimals. Then, line up the decimal points and compare digits from left to right.

Example

Place in order from least to greatest: 0.21, $\frac{1}{5}$, 19%

Convert the fraction $\frac{1}{5}$ to a decimal: $1 \div 5 = 0.2$. Convert 19% to a decimal: $19\% = 0.19$. Line up the decimal points and add trailing zeros as needed:

$$0.21$$
$$0.20$$
$$0.19$$

The value 0.19 has a 1 in the tenths place, while 0.21 and 0.20 each have a 2 in the tenths place, so 0.19 is the smallest value. 0.20 has a 0 in the hundredths place and 0.21 has a 1 in the hundredths place, so 0.20 is less than 0.21. Therefore, $0.19 < 0.20 < 0.21$, and $19\% < \frac{1}{5} < 0.21$.

Example

Place in order from least to greatest: 925.1%, $9\frac{1}{4}$, 9.2501, $9\frac{3}{10}$, 929%

$925.1\% = 9.251$, $9\frac{1}{4} = 9.25$, $9\frac{3}{10} = 9.3$, and $929\% = 9.29$. Line up the decimal points, add trailing zeros, and compare:

$$9.2510$$
$$9.2500$$
$$9.2501$$
$$9.3000$$
$$9.2900$$

Each number has a 9 in the ones place, so we move to the tenths place. Every number has a 2 in the tenths place except 9.3000, which has a 3, so 9.300 is the greatest number. Now look at the hundredths digits of the remaining numbers. The first three numbers have a 5 in the hundredths place, while 9.2900 has a 9 in the hundredths place. Because $9 > 5$, 9.2900 is the next largest number. Now move to the thousandths place. The remaining numbers all have a 0 in the thousandths place except for 9.2510, which has a 1, so it is the next largest number. That leaves 9.2500 and 9.2501. Because 9.2501 has a 1 in the ten thousandths place and 9.2500 has a 0 in the ten thousandths place, 9.2501 is the larger number. $9.2500 < 9.2501 < 9.2510 < 9.2900 < 9.300$, and $9\frac{1}{4} < 9.2501 < 925.1\% < 929\% < 9\frac{3}{10}$.

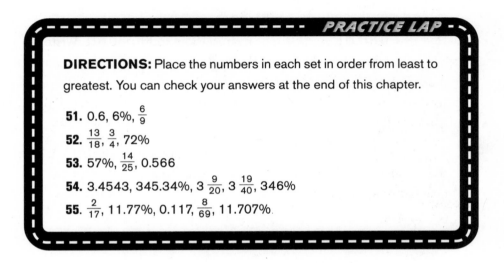

PRACTICE LAP

DIRECTIONS: Place the numbers in each set in order from least to greatest. You can check your answers at the end of this chapter.

51. 0.6, 6%, $\frac{6}{9}$

52. $\frac{13}{18}$, $\frac{3}{4}$, 72%

53. 57%, $\frac{14}{25}$, 0.566

54. 3.4543, 345.34%, $3\frac{9}{20}$, $3\frac{19}{40}$, 346%

55. $\frac{2}{17}$, 11.77%, 0.117, $\frac{8}{69}$, 11.707%

Given a fraction, decimal, or percent, we can now turn any one of the three into the other two. Having learned how to perform every major operation with fractions, decimals, and percents, you're ready for the posttest.

ANSWERS

1. $3 \div 12 = 0.25$. $\frac{3}{12} = 0.25$

2. $5 \div 8 = 0.625$. $\frac{5}{8} = 0.625$

3. $9 \div 16 = 0.5625$. $\frac{9}{16} = 0.5625$

4. Place 2 in the hundredths place: $\frac{2}{100} = 0.02$.

5. Because 761 is three digits and not all three can be placed in the ten thousandths place, use the thousandths and hundredths places to place the digits 7 and 6, respectively: $\frac{761}{10,000} = 0.0761$.

6. $6 \div 9 = 0.6666. \ldots \frac{6}{9}$, to the nearest thousandth, is 0.667.

7. $4 \div 11 = 0.363636. \ldots \frac{4}{11}$, to the nearest thousandth, is 0.364.

8. $5 \div 7 = 0.714285. \ldots \frac{5}{7}$, to the nearest thousandth, is 0.714.

9. $7 \div 12 = 0.5833333. \ldots \frac{7}{12}$, to the nearest thousandth, is 0.583.

10. $98 \div 762 = 0.1286089. \ldots \frac{98}{762}$, to the nearest thousandth, is 0.129.

11. $5 \div 32 = 0.15625. \frac{5}{32} = 0.15625$. Write the whole number, 10, to the left of the decimal point. $10\frac{5}{32} = 10.15625$.

12. $11 \div 18 = 0.611111. \ldots \frac{11}{18}$, to the nearest thousandth, is 0.611. Write the whole number, 7, to the left of the decimal point. $7\frac{11}{18}$, to the nearest thousandth, is 7.611.

13. $19 \div 20 = 0.95. \frac{19}{20} = 0.95$. Write the whole number, 43, to the left of the decimal point. $43\frac{19}{20} = 43.95$.

14. $3 \div 26 = 0.1153846. \ldots \frac{3}{26}$, to the nearest thousandth, is 0.115. Write the whole number, 942, to the left of the decimal point. $942\frac{3}{26}$, to the nearest thousandth, is 942.115.

15. $0.47 = 47\%$

16. $0.9 = 90\%$

17. $0.133 = 13.3\%$

18. $1.52 = 152\%$

19. $9.01 = 901\%$

20. $49.2227 = 4,922.27\%$

21. $8 \div 10 = 0.8 = 80\%. \frac{8}{10} = 80\%$

22. $22 \div 64 = 0.34375 = 34.375\%. \frac{22}{64} = 34.375\%$

23. $3 \div 19 = 0.157894 \ldots \frac{3}{19}$, to the nearest thousandth, is 0.158; 0.158 $= 15.8\%. \frac{3}{19}$ is approximately 15.8%.

24. $17 \div 30 = 0.566666 \ldots \frac{17}{30}$, to the nearest thousandth, is 0.567. Write the whole number, 7, to the left of the decimal point. $7\frac{17}{30}$ is approximately 7.567, which equals 756.7%.

25. $53 \div 80 = 0.6625$. Write the whole number, 87, to the left of the decimal point. $87\frac{53}{80} = 87.6625 = 8,766.25\%$.

26. 0.33 is "thirty-three hundredths," or $\frac{33}{100}$.

27. 0.2 is "two tenths," or $\frac{2}{10}$. The greatest common factor of 2 and 10 is 2, so both the numerator and denominator of $\frac{2}{10}$ can be divided by 2. $\frac{2}{10}$ $= \frac{1}{5}$

28. 0.655 is "six hundred fifty-five thousandths," or $\frac{655}{1,000}$. The greatest common factor of 655 and 1,000 is 5, so both the numerator and denominator of $\frac{655}{1,000}$ can be divided by 5. $\frac{655}{1,000} = \frac{131}{200}$.

29. 4.07 is "four and seven hundredths," or $4\frac{7}{100}$.

30. 19.375 is "nineteen and three hundred seventy-five thousandths," or 19 $\frac{375}{1,000}$. The greatest common factor of 375 and 1,000 is 125, so both the numerator and denominator of $\frac{375}{1,000}$ can be divided by 125. $\frac{375}{1,000} = \frac{3}{8}$. $19.375 = 19\frac{3}{8}$.

31. 73% = 0.73

32. 1% = 0.01

33. 63.24% = 0.6324

34. 252.1% = 2.521

35. 0.039% = 0.00039

36. 9% = 0.09, which is "nine hundredths," or $\frac{9}{100}$.

37. 26% = 0.26, which is "twenty-six hundredths," or $\frac{26}{100}$. The greatest common factor of 26 and 100 is 2, so both the numerator and denominator of $\frac{26}{100}$ can be divided by 2. $\frac{26}{100} = \frac{13}{50}$.

38. 88.1% = 0.881, which is "eight hundred eighty-one thousandths," or $\frac{881}{1,000}$.

39. 429% = 4.29, which is "four and twenty-nine hundredths," or $4\frac{29}{100}$.

40. 0.667% = 0.00667, which is "six hundred sixty-seven hundred-thousandths," or $\frac{667}{100,000}$.

41. 5.96% = 0.0596. Because the tenths digit of 0.596 (5) is greater than the tenths digit of 0.0596 (0), 0.596 > 0.0596 and 0.596 > 5.96%.

42. $\frac{17}{20}$ = 17 ÷ 20 = 0.85. Because the tenths digit of 0.85 (8) is greater than the tenths digit of 0.56 (5), 0.85 > 0.59 and $\frac{17}{20}$ > 0.59.

43. 72% = 0.72. $\frac{3}{4}$ = 3 ÷ 4 = 0.75. Because the hundredths digit of 0.72 (2) is less than the hundredths digit of 0.75 (5), 0.72 < 0.75 and 72% < $\frac{3}{4}$.

44. $\frac{21}{70}$ = 21 ÷ 70 = 0.3. 30% = 0.3. Because each digit in each number is the same, 0.3 = 0.3 and $\frac{21}{70}$ = 30%.

45. $\frac{2}{9}$ = 2 ÷ 9 = 0.2222. . ., or 0.222 to the nearest thousandth. Because the hundredths digit of 4.29 (9) is greater than the hundredths digit of 4.222 (2), 4.29 > 4.222 and 4.29 > $4\frac{2}{9}$.

46. $\frac{11}{21}$ = 11 ÷ 21 = 0.5238095. . ., or 0.524 to the nearest thousandth. However, the decimal 0.5238 has a digit in the thousandths place, and the ten thousandths place, so we must round the result of 11 ÷ 21 to the hundred thousandths place. 0.5238095 rounded to the hundred thousandths place is 0.52381. Because the hundred thousandths digit of 0.52380 (0) is less than the hundred thousandths digit of 0.52381 (1), 0.5238 < 0.52381 and 0.5238 < $\frac{11}{21}$.

47. 0.1% = 0.001. Because each digit in each number is the same, 0.001 = 0.001 and 0.001 = 0.1%.

48. $\frac{5}{7}$ = 5 ÷ 7 = 0.7142857. . ., or 0.714 to the nearest thousandth. However, the decimal 0.143 has a digit in the thousandths place, so we must round the result of 5 ÷ 7 to the ten thousandths place. 0. 7142857 rounded to the ten thousandths place is 0.7143. Because the tenths digit of 7.7143 (7) is greater than the tenths digit of 7.143 (1), 7.7143 > 7.143 and $7\frac{5}{7}$ > 7.143.

49. 29.45% = 0.2945. Because the thousands digit of 00.2945 (0) is less than the thousands digit of 2,945 (2), 00.2945 < 2,945 and 29.45% < 2,945.

50. $\frac{2}{3}$ = 2 ÷ 3 = 0.6666666, or 0.667 to the nearest thousandth. 67% = 0.67. Because the hundredths digit of 0.667 (6) is less than the hundredths digit 0.67 (7), 0.667 < 0.67 and $\frac{2}{3}$ < 67%.

51. 6% = 0.06, $\frac{6}{9}$ = 0.6666. . ., or 0.667 to the nearest thousandth. Line up the decimal points and add trailing zeros:

$$0.060$$

$$0.600$$

$$0.667$$

The tenths digit of 0.060 is 0 while the tenths digits of 0.600 and 0.667 are 6, so 0.060 is the smallest number. The hundredths digit of 0.600 is 0 while the hundredths digit of 0.667 is 6, so 0.600 is less than 0.667. 0.06 < 0.600 < 0.667, so 6% < 0.6 < $\frac{6}{9}$.

52. $\frac{13}{18} = 0.722222\ldots$, or 0.722 to the nearest thousandth. $\frac{3}{4} = 0.75$ and 72% $= 0.72$. Line up the decimal points and add trailing zeros:

0.722

0.750

0.720

The tenths digits of all three numbers are the same. The hundredths digit of 0.750 (5) is greater than the hundredths digits of 0.720 and 0.722 (2), so 0.750 is the largest number. The thousandths digit of 0.722 (2) is larger than the thousandths digit of 0.720 (0), so 0.722 is greater than 0.720. 0.720 < 0.722 < 0.750, so 72% $< \frac{13}{18} < \frac{3}{4}$.

53. 57% $= 0.57$, $\frac{14}{25} = 0.56$. Line up the decimal points and add trailing zeros:

0.570

0.560

0.566

The tenths digits of all three numbers are the same. The hundredths digit of 0.570 (7) is greater than the hundredths digits of 0.560 and 0.566 (6), so 0.570 is the largest number. The thousandths digit of 0.566 (6) is larger than the thousandths digit of 0.560 (0), so 0.566 is greater than 0.560. 0.560 < 0.566 < 0.570, so $\frac{14}{25} < 0.566 < 57\%$.

54. 345.34% $= 3.4534$, $3\frac{9}{20} = 3.45$, $3\frac{19}{40} = 3.475$, 346% $= 3.46$. Line up the decimal points and add trailing zeros:

3.4543

3.4534

3.4500

3.4750

3.4600

The ones and tenths digits of all three numbers are the same. The hundredths digit of the first three numbers (5) is less than the hundredths digits of 3.4750 (7) and 3.4600 (6), so 3.4750 is the largest number, followed by 3.4600. Now compare the thousandths digits of the first three numbers. The thousandths digit of 3.4543 (4) is larger than the thousandths digits of 3.4534 (3) and 3.4500 (0), so 3.4543 is greater than

3.4534, and 3.4534 is greater than 3.4500. $3.4500 < 3.4534 < 3.4543 < 3.4600 < 3.4750$, so $3\frac{9}{20} < 345.34\% < 3.4543 < 346\% < 3\frac{19}{40}$.

55. $\frac{2}{17} = 0.117647\ldots$, or 0.11765 to the nearest hundred thousandth. $11.77\% = 0.1177$, $\frac{8}{69} = 0.115942\ldots$, or 0.11594 to the nearest hundred thousandth, $11.707\% = 0.11707$. Line up the decimal points and add trailing zeros:

$$0.11765$$
$$0.11770$$
$$0.11700$$
$$0.11594$$
$$0.11707$$

The ones, tenths, and hundredths digits of all five numbers are the same. The thousandths digit of 0.11594 (5) is less than the thousandths digits of the other four numbers (7), so 0.11594 is the smallest number. Compare the ten thousandths digits of the remaining numbers. The number 0.11770 has the greatest ten thousandths digit (7), followed by 0.11765 (6), which makes these two numbers the largest numbers. Now look at the hundred thousandths digit of the two remaining numbers, 0.11700 and 0.11707. The hundred thousandths digit of 0.11700 is 0 and the hundred thousandths digit of 0.11707 is 7, which means that 0.11700 is less than 0.11707. $0.11594 < 0.11700 < 0.11707 < 0.11765 < 0.11770$, and $\frac{8}{69} < 0.117 < 11.707\% < \frac{2}{17} < 11.77\%$.

Posttest

Now that you have read all the chapters in this book, you are ready to take the posttest. Like the pretest, the posttest has 50 questions that are presented in the same order that the topics in which they cover are presented in this book. The questions are very similar to the pretest, so that you can see your progress and improvement since taking the pretest.

After completing the posttest, check your answers. Explanations are provided for every answer, along with the chapter in which the skill being tested is taught. Compare your score on the posttest to your score on the pretest. Did you improve? Did you learn from any mistakes you made in the pretest? The posttest can help you identify which areas you have mastered and which areas you need to practice. Return to the chapters that cover the topics that are tough for you until those topics become your strengths. You'll be able to ace the pretest, the posttest, and any other test on fractions, decimals, and percents.

Answer the questions that follow. Reduce all fractions to their simplest form and round all decimals to the nearest thousandth.

1. What fraction of the rectangle below is shaded?

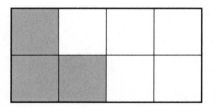

2. Find the greatest common factor of 36 and 60.

3. Find the least common denominator for $\frac{1}{6}$ and $\frac{8}{15}$.

4. Which fraction is greater, $\frac{6}{8}$ or $\frac{17}{20}$?

5. $\frac{1}{14} + \frac{8}{14} =$

6. $\frac{7}{10} + \frac{11}{15} =$

7. $\frac{7}{9} - \frac{6}{9} =$

8. $\frac{11}{12} - \frac{5}{8} =$

9. $\frac{5}{7} \times \frac{7}{20} =$

10. $\frac{3}{4} \div \frac{8}{3} =$

11. Convert $3\frac{1}{3}$ to a mixed number.

12. Convert $12\frac{2}{11}$ to an improper fraction.

13. $8\frac{2}{9} + 22\frac{5}{18} =$

14. $10\frac{5}{7} - 9\frac{6}{7} =$

15. $5\frac{4}{13} \times 5\frac{3}{5} =$

16. $5\frac{5}{2} \div 3\frac{3}{4} =$

17. Simplify this complex fraction: $\frac{21}{\left(\frac{1}{7}\right)}$.

18. Write this division sentence as a complex fraction: $\frac{9}{17} \div \frac{14}{23}$.

19. Write this ratio in simplest form: 52:13.

20. The ratio of red squares to blue squares in a quilt is 4:7. If there are 154 squares in the quilt, how many of them are red?

21. In the number 40.8761293, in what place is the number 1?

22. What is the value of the digit 6 in 1.2991633?

23. Write 504.066 in words.

24. Write "twenty-two thousand six hundred seventeen and three hundred ninety-nine ten thousandths" in digits.

25. Round 88,745.159 to the nearest hundred.

26. Which decimal is smaller, 774.316794 or 774.3167499?

27. 745.28 + 0.0441 =

28. 5,622 − 84.2551 =

29. $0.436 \times 7.06 =$

30. $2,466.6192 \div 7.2 =$

31. Write 562.74% as a decimal.

32. What is 4% of 76.65?

33. Write 1.0883 as a percent.

34. What percent of 41 is 95?

35. What is 214% of 26.1?

36. What percent of 835 is 3.5?

37. Find the percent increase from 165 to 192.225.

38. If the value 693 increases by 0.3%, what is the new value?

39. A value is increased by 82.7% to 628.488. What was the original value?

40. Find the percent decrease from 14 to −5.6.

41. If the value 57 decreases by 37.5%, what is the new value?

42. A value is decreased by 0.07% to 1,249.125. What was the original value?

43. Rori makes a sale of $30,000 and receives a 6.25% commission, or 6.25% of the total sale. What was Rori's commission?

44. Jonathan gains $141.81 in interest after keeping his principal in a bank for 9 months at a rate of 5.8%. What was his principal?

45. Write $366\frac{19}{24}$ as a decimal.

46. Write $5\frac{59}{128}$ as a percent.

47. Write 0.018 as a fraction.

48. Write 29.41% as a fraction.

49. Which is greater, 81.57% or $\frac{31}{38}$?

50. Which of the following numbers are equal?
    71.88%, $\frac{69}{96}$, 0.7188

ANSWERS

1. The rectangle is divided into eight parts, so the denominator of the fraction will be 8. There are 3 parts of the rectangle that are shaded, so the numerator of the fraction will be 3. Because 3 out of 8 parts are shaded, $\frac{3}{8}$ of the rectangle is shaded. For more on this concept, see Chapter 3.

2. The factors of 36 are 1, 2, 3, 4, 6, 9, **12**, 18, and 36, and the factors of 60 are 1, 2, 3, 4, 5, 6, 10, **12**, 15, 20, 30, and 60. The greatest common factor (the largest number that is a factor of both 36 and 60) is 12. For more on this concept, see Chapter 3.

3. To find the least common denominator for these fractions, we must find the least common multiple of 6 and 15. List the multiples of each number:

$$6: 6, 12, 18, 24, \mathbf{30}, 36, \ldots$$
$$15: 15, \mathbf{30}, 45, 60, 75, \ldots$$

The least common multiple of 6 and 15 is 30, so the least common denominator for $\frac{1}{6}$ and $\frac{8}{15}$ is 30. For more on this concept, see Chapter 3.

4. In order to compare these fractions, we must find common denominators for them, so we must find the least common multiple of 8 and 20. List the multiples of each number:

$$8: 8, 16, 24, 32, \mathbf{40}, 48, 56, \ldots$$
$$20: 20, \mathbf{40}, 60, 80, 100, \ldots$$

The least common multiple of 8 and 20 is 40. Convert both fractions to a number over 40. Because $\frac{40}{8} = 5$, the new denominator of the fraction $\frac{6}{8}$ is five times larger. Therefore, the new numerator must also be five times larger, so that the value of the fraction does not change: $6 \times 5 = 30$. $\frac{6}{8} = \frac{30}{40}$. Because $\frac{40}{20} = 2$, the new denominator of the fraction $\frac{17}{20}$ is two times larger. Multiply the numerator of the fraction by two: $17 \times 2 = 34$. $\frac{17}{20} = \frac{34}{40}$. Now that we have two fractions with common denominators, we can compare their numerators. Because 34 is greater than 30, $\frac{34}{40} > \frac{30}{40}$ and $\frac{17}{20} > \frac{6}{8}$. For more on this concept, see Chapter 3.

5. These fractions are like fractions, because they have the same denominator. To add two like fractions, add the numerators of the fractions:.

$1 + 8 = 9$. The denominator of our answer is the same as the denominator of the two fractions that we are adding. Both fractions have a denominator of 14, so the denominator of our answer will be 14: $\frac{1}{14} + \frac{8}{14} = \frac{9}{14}$. For more on this concept, see Chapter 4.

6. Because these fractions are unlike, we must find common denominators for them before adding. First, we find the least common multiple of 10 and 15:

$$10: 10, 20, \mathbf{30}, 40, 50, \ldots$$

$$15: 15, \mathbf{30}, 45, 60, 75, \ldots$$

The least common multiple of 10 and 15 is 30. Convert both fractions to a number over 30. Because $\frac{30}{10} = 3$, the new denominator of the fraction $\frac{7}{10}$ is three times larger. Therefore, the new numerator must also be three times larger, so that the value of the fraction does not change: $7 \times 3 = 21$. $\frac{7}{10} = \frac{21}{30}$. Because $\frac{30}{15} = 2$, the new denominator of the fraction $\frac{11}{15}$ is two times larger. Multiply the numerator of the fraction by two: $11 \times 2 = 22$. $\frac{11}{15} = \frac{22}{30}$. Now that we have like fractions, we can add the numerators: $21 + 22 = 43$. Because the denominator of the fractions we are adding is 30, the denominator of our answer is 30. $\frac{21}{30} + \frac{22}{30} = \frac{43}{30}$. Convert the improper fraction to a mixed number by dividing the numerator by the denominator. 43 divided by 30 is 1 with 13 left over. We express the remainder as a fraction. Because the improper fraction has a denominator of 30, our remainder has a denominator of 30; 43 divided by 30 is $1\frac{13}{30}$. For more on this concept, see Chapter 4.

7. These fractions are like fractions, so we can subtract the numerator of the second fraction from the numerator of the first fraction: $7 - 6 = 1$. Because the denominator of both fractions is 9, the denominator of our answer is 9: $\frac{7}{9} - \frac{6}{9} = \frac{1}{9}$. For more on this concept, see Chapter 4.

8. Because these fractions are unlike, we must find common denominators for them before subtracting. First, we find the least common multiple of 12 and 8:

$$12: 12, \mathbf{24}, 36, 48, 60, \ldots$$

$$8: 8, 16, \mathbf{24}, 32, 40, 48, \ldots$$

The least common multiple of 12 and 8 is 24. Convert both fractions to a number over 24. Because $\frac{24}{12} = 2$, the new denominator of the fraction $\frac{11}{12}$ is two times larger. Therefore, the new numerator must also be two times larger, so that the value of the fraction does not change: $11 \times 2 = 22$. $\frac{11}{12} = \frac{22}{24}$. Because $\frac{24}{8} = 3$, the new denominator of the fraction $\frac{5}{8}$ is three times larger. Multiply the numerator of the fraction by three: $5 \times 3 = 15$. $\frac{5}{8} = \frac{15}{24}$. Now that we have like fractions, we can subtract the second numerator from the first numerator: $22 - 15 = 7$. Because the denominator of the fractions is 24, the denominator of our answer is 24. $\frac{22}{24} - \frac{15}{24} = \frac{7}{24}$. For more on this concept, see Chapter 4.

9. The product of two fractions is equal to the product of the numerators over the product of the denominators. Before multiplying, we can simplify this multiplication problem. Divide the denominator of the first fraction and the numerator of the second fraction by 7. Then, divide the numerator of the first fraction and the denominator of the second fraction by 5. The problem becomes $\frac{1}{1} \times \frac{1}{4}$. Multiply the numerators: $1 \times 1 = 1$. Multiply the denominators: $1 \times 4 = 4$. $\frac{1}{1} \times \frac{1}{4} = \frac{1}{4}$. For more on this concept, see Chapter 4.

10. To divide a fraction by a fraction, first find the reciprocal of the divisor (the fraction by which you are dividing). In the number sentence $\frac{3}{4} \div \frac{8}{3}$, $\frac{8}{3}$ is the divisor. The reciprocal of a number can be found be switching the numerator with the denominator. The reciprocal of $\frac{8}{3}$ is $\frac{3}{8}$. Now, multiply the dividend $(\frac{3}{4})$ by the reciprocal of the divisor: $\frac{3}{4} \times \frac{3}{8}$. Multiply the numerators: $3 \times 3 = 9$. Multiply the denominators: $4 \times 8 = 32$. $\frac{3}{4} \times \frac{3}{8} = \frac{9}{32}$, so $\frac{3}{4} \div \frac{8}{3} = \frac{9}{32}$. For more on this concept, see Chapter 4.

11. To convert an improper fraction to a mixed number, divide the numerator by the denominator. 31 divided by 3 is 10 with 1 left over. We express the remainder as a fraction. Because the improper fraction has a denominator of 3, our remainder has a denominator of 3. 31 divided by 3 is $10\frac{1}{3}$. For more on this concept, see Chapter 5.

12. To convert a mixed number to an improper fraction, we begin by multiplying the whole number, 12, by the denominator of the fraction, 11: $12 \times 11 = 132$. Next, add to that product the numerator of the fraction: $132 + 2 = 134$. Finally, put that sum over the denominator of the fraction. $12\frac{2}{11} = \frac{134}{11}$. For more on this concept, see Chapter 5.

13. To add two mixed numbers, we can either add the whole number parts and then add the fractional parts, or, we can convert both mixed numbers to improper fractions and then add. It may be easier to use the second method if the sum of the fractional parts will be greater than 1. However, because $\frac{2}{9}$ and $\frac{5}{18}$ are both less than $\frac{1}{2}$, their sum will be less than 1. We will add the whole number parts and then add the fractional parts. $8 + 22 = 30$. To add $\frac{2}{9}$ and $\frac{5}{18}$, we must first find common denominators. List the multiples of 9 and 18:

$$9: 9, \mathbf{18}, 36, 45, \ldots$$
$$18: \mathbf{18}, 36, 54, \ldots$$

The least common multiple of 9 and 18 is 18. Convert $\frac{2}{9}$ to a number over 18. Because $\frac{18}{9} = 2$, the new denominator of the fraction $\frac{2}{9}$ is two times larger. Therefore, the new numerator must also be two times larger, so that the value of the fraction does not change: $2 \times 2 = 4$. $\frac{2}{9} = \frac{4}{18}$. Now that we have like fractions, we can add the numerators: $4 + 5 = 9$. Because the denominator of the fractions we are adding is 18, the denominator of our answer is 18. $\frac{4}{18} + \frac{5}{18} = \frac{9}{18}$. We can reduce this fraction. The greatest common factor of 9 and 18 is 9, so divide the numerator and denominator of the fraction by 9. $\frac{9}{9} = 1$ and $\frac{18}{9} = 2$. $\frac{9}{18} = \frac{1}{2}$. $8\frac{2}{9} + 22\frac{5}{18} = 30\frac{9}{18} = 30\frac{1}{2}$. For more on this concept, see Chapter 5.

14. The fraction parts of the mixed numbers have the same denominator, but the fraction being subtracted ($\frac{6}{7}$) is bigger than the fraction from which it is being subtracted ($\frac{5}{7}$). In a situation like this, it is easier to convert both mixed numbers to improper fractions before subtracting. Multiply the whole number part of the mixed number by the denominator of the fraction, and then add that product to the numerator of the fraction. $10 \times 7 = 70$, $70 + 5 = 75$, so $10\frac{5}{7} = \frac{75}{7}$; $9 \times 7 = 63$, $63 + 6 = 69$, so $9\frac{6}{7} = \frac{69}{7}$. Because we have like fractions, subtract the numerator of $\frac{69}{7}$ from the numerator of $\frac{75}{7}$ and keep the denominator. $75 - 69 = 6$, so $\frac{75}{7} - \frac{69}{7} = \frac{6}{7}$, which means that $10\frac{5}{7} - 9\frac{6}{7} = \frac{6}{7}$. For more on this concept, see Chapter 5.

15. Begin by converting both mixed numbers to improper fractions. Multiply the whole number by the denominator of the fraction, and then add that product to the numerator of the fraction. $5 \times 13 = 65$, $65 + 4$

234 Express Review Guides: FRACTIONS, PERCENTAGES, & DECIMALS

$= 69$, so $5\frac{4}{13} = \frac{69}{13}$. $5 \times 5 = 25$, $25 + 3 = 28$, so $5\frac{3}{5} = \frac{28}{5}$. Now we have $\frac{69}{13} \times \frac{28}{5}$. The product of two fractions is equal to the product of the numerators over the product of the denominators. Multiply the numerators: $69 \times 28 = 1{,}932$. Multiply the denominators: $13 \times 5 = 65$. $\frac{69}{13}\frac{28}{5}$ $= \frac{1{,}932}{65} \cdot \frac{1{,}932}{65} = 29$ with 47 left over. We express the remainder as a fraction. Because the improper fraction has a denominator of 65, our remainder has a denominator of 65; 1,932 divided by 65 is $29\frac{47}{65}$. $5\frac{4}{13} \times$ $5\frac{3}{5} = 29\frac{47}{65}$. For more on this concept, see Chapter 5.

16. Begin by converting both mixed numbers to improper fractions. Multiply the whole number by the denominator of the fraction, and then add that product to the numerator of the fraction. $5 \times 2 = 10$, $10 + 5 =$ 15, so $5\frac{5}{2} = \frac{15}{2}$. $3 \times 4 = 12$, $12 + 3 = 15$, so $3\frac{3}{4} = \frac{15}{4}$. Now we have $\frac{15}{2} \div$ $\frac{15}{4}$. To divide two fractions, we take the reciprocal of the divisor and then multiply. To find the reciprocal of the divisor, $\frac{15}{4}$, we switch the numerator and the denominator. The reciprocal of $\frac{15}{4}$ is $\frac{4}{15}$. The problem is now $\frac{15}{2} \times \frac{4}{15}$. Divide the numerator of the first fraction and the denominator of the second fraction by 15. Then, divide the denominator of the first fraction and the numerator of the second fraction by 2. The problem becomes $\frac{1}{1} \times \frac{2}{1}$. Multiply the numerators: $1 \times 2 = 2$. Multiply the denominators: $1 \times 1 = 1$. $\frac{1}{1} \div \frac{2}{1} = \frac{2}{1} = 2$. $5\frac{5}{2} \div 3\frac{3}{4} = 2$. For more on this concept, see Chapter 5.

17. A complex fraction, like any fraction, represents a division sentence: $\frac{21}{(\frac{1}{7})}$ $= 21 \div \frac{1}{7}$. To divide two fractions, we take the reciprocal of the divisor and then multiply. The reciprocal of $\frac{1}{7}$ is $\frac{7}{1}$, or 7. The problem is now 21×7, which is equal to 147. For more on this concept, see Chapter 6.

18. The first fraction, the dividend, becomes the numerator of the fraction and the second fraction, the divisor, becomes the denominator of the fraction. $\frac{9}{17} \div \frac{14}{23} = \frac{(\frac{9}{17})}{(\frac{14}{23})}$. For more on this concept, see Chapter 6.

19. Reducing a ratio is just like reducing a fraction. Instead of finding the greatest common factor of a numerator and a denominator, find the greatest common factor of the two numbers in the ratio. The factors of 52 are 1, 2, 4, **13**, 26, and 52, and the factors of 13 are 1 and **13**. The greatest common factor is 13, so divide each number by 13: $\frac{52}{13} = 4$ and $\frac{13}{13} =$ 1. The ratio 52:13 reduces to 4:1. For more on this concept, see Chapter 6.

20. We need to find a ratio that compares the number of red squares to the total number of squares. If the ratio of red squares to blue squares is 4:7, then the ratio of red squares to total squares is 4 to 4 + 7 = 11, or 4:11. Write the ratio as a fraction: 4:11 = $\frac{4}{11}$. Set up a proportion. If 4 squares of 11 total squares are red, then x squares of 154 total squares are red: $\frac{4}{11} = \frac{x}{154}$. Cross multiply: $(11)(x) = (4)(154)$, $11x = 616$. Divide both sides of the equation by 11: $\frac{11x}{11} = \frac{616}{11}$, $x = 56$. If there are 154 total squares, then 56 of them are red. For more on this concept, see Chapter 6.

21. In the number 40.8761293, the number 1 is four digits to the right of the decimal point. The fourth place to the right of the decimal point is the ten thousandths place. For more on this concept, see Chapter 7.

22. The 6 is in the hundred thousandths place, so it has a value of $6 \times 0.00001 = 0.00006$. For more on this concept, see Chapter 7.

23. Five hundred four and sixty-six thousandths. For more on this concept, see Chapter 7.

24. 22,617.0399. For more on this concept, see Chapter 7.

25. The digit to the immediate right of the hundreds place is the digit in the tens place, 4. Because 4 is less than 5, we round down. The hundreds digit remains 7, and the digits to the right become zero. 88,745.159 to the nearest hundred is 88,700.000, or just 88,700. For more on this concept, see Chapter 7.

26. To compare two decimals, line up the decimal points, place trailing zeros, and compare corresponding digits from left to right:

774.3167940
774.3167499

Both numbers have the same digits in the hundreds, tens, ones, tenths, hundredths, thousandths, and ten thousandths places, so we move to the hundred thousandths place. The first number has a 9 in the hundred thousandths place and the second number has a 4 in the hundred thousandths place. Because 9 is greater than 4, 774.3167499 is smaller than 774.3167940. For more on this concept, see Chapter 7.

27. To add two decimals, write the problem vertically and place trailing zeros on the number with fewer places to the right of the decimal point:

745.2800
+ 0.0441

Add column by column and carry the decimal point down into your answer.

745.2800
+ 0.0441
745.3241

For more on this concept, see Chapter 8.

28. To subtract a decimal from a whole number, begin by writing the problem vertically, lining up the decimal points. Put a decimal on the end of the whole number, and place trailing zeros next to it:

5,622.0000
− 84.2551

Subtract column by column and carry the decimal point down into your answer.

5,622.0000
− 84.2551
5,537.7449

For more on this concept, see Chapter 8.

29. 0.436 has three digits to the right of the decimal point and 7.06 has two digits to the right of the decimal point, so our answer will have five (2 + 3) digits to the right of the decimal point:

0.436
× 7.06
2616
000
3052
3.07816

For more on this concept, see Chapter 8.

30. The divisor has a decimal point and one digit to the left of the decimal point. Shift both the decimal point in the divisor and the decimal point in the dividend one place to the right, and then divide.

$$
\begin{array}{r}
342.586 \\
72\overline{)24{,}666.192} \\
216 \\
\hline
306 \\
288 \\
\hline
186 \\
144 \\
\hline
421 \\
360 \\
\hline
619 \\
576 \\
\hline
432 \\
432 \\
\hline
0
\end{array}
$$

For more on this concept, see Chapter 8.

31. To write a percent as a decimal, move the decimal point two places to the left. 562.74% = 5.6274. For more on this concept, see Chapter 9.

32. First, write 4% as a decimal: 4% = 0.04. Then, multiply the decimal by the number: $0.04 \times 76.65 = 3.066$. For more on this concept, see Chapter 9.

33. To write a decimal as a percent, move the decimal point two places to the right. 1.0883 = 108.83%. For more on this concept, see Chapter 9.

34. To find what percent of 41 is 95, divide 95 by 41: $\frac{95}{41} = 2.317073...$, or 2.317 to the nearest thousandth. To convert a decimal to a percent, move the decimal point two places to the right and add the percent sign. 2.317 = 231.7%. For more on this concept, see Chapter 9.

35. We follow the same steps we used when working with percents that are less than 100. Write the percent as a decimal and multiply. Move the decimal point two places to the left: 214% = 2.14. $2.14 \times 26.1 = 55.854$. For more on this concept, see Chapter 9.

36. Divide 835 by 3.5: $3.5 \div 835 = 0.0041916...$, or 0.004 to the nearest thousandth. Write the decimal as a percent by moving the decimal point two places to the right. 0.004 = 0.4%; 3.5 is about 0.4% of 835. For more on this concept, see Chapter 9.

37. The original value is 165 and the new value is 192.225. Subtract the original value from the new value: 192.225 − 165 = 27.225. Next, divide the difference by the original value: 27.225 ÷ 165 = 0.165. Now, write the decimal as a percent by moving the decimal point two places to the right: 0.165 = 16.5%. For more on this concept, see Chapter 10.

38. The percent increase is 0.3%, or 0.003. Add 1 and multiply by the original value: 693 × 1.003 = 695.079. For more on this concept, see Chapter 10.

39. Write the percent increase as a decimal: 82.7% = 0.827. Add 1 to the percent increase and divide the new value by that sum: 628.488 ÷ 1.827 = 344. For more on this concept, see Chapter 10.

40. Subtract the new value from the original value and divide by the original value: 14 − (−5.6) = 19.6, 19.6 ÷ 14 = 1.4. Change the number to a percent by multiplying by 100, which moves the decimal two places to the right. 1.4 = 140%. For more on this concept, see Chapter 10.

41. The original value is 57 and the percent decrease is 37.5%, or 0.375. **(new value) = (1 − percent decrease) × (original value)**. Substitute the values into the formula: (1 − 0.375) × 57 = 0.625 × 57 = 35.625. For more on this concept, see Chapter 10.

42. Write the percent decrease as a decimal: 0.0007. **(original value) = (new value) ÷ (1 − percent decrease)**. Subtract the percent decrease from 1 and divide the new value by that difference: 1,249.125 ÷ (1 − .0007) = 1,249.125 ÷ 0.9993 = 1,250. For more on this concept, see Chapter 10.

43. To find the amount of a commission on a sale, convert the commission percent to a decimal, and then multiply it by the total sale: 6.25% = 0.0625, 0.0625 × $30,000 = $1,875. For more on this concept, see Chapter 10.

44. The formula for finding principal is $p = I \div rt$, where p is principal, I is interest, r is the rate, and t is the time in years. The time is given in months, so it must be converted to years. There are 12 months in a year, so 9 months is equal to $\frac{9}{12}$ = 0.75 years. Write the rate as a decimal and substitute the values into the formula. 5.8% = 0.058. $p = $141.81 ÷ (0.058 × 0.75) = $3,260. For more on this concept, see Chapter 10.

45. To write a mixed number as a decimal, begin by writing the whole number part, 366, to the left of the decimal point. Then, write the fraction part as a decimal by dividing the numerator by the denominator. 19 ÷ 24 = 0.79166666. . ., or 0.792 to the nearest thousandth, so $366\frac{19}{24}$ = 366.792, to the nearest thousandth. For more on this concept, see Chapter 11.

46. To write a fraction as a percent, first write the fraction as a decimal. Write the whole number part, 5, to the left of the decimal point, and write the fraction part as a decimal by dividing the numerator by the denominator 59 ÷ 128 = 0.4609375, or 0.461 to the nearest thousandth. $5\frac{59}{128}$ is equal to 5.461, to the nearest thousandth. Now, write the decimal as a percent by moving the decimal point two places to the right and adding the percent symbol. 5.461 = 546.1%. For more on this concept, see Chapter 11.

47. To write a decimal as a fraction, first read the name of the decimal aloud: 0.018 is "eighteen thousandths." The first part of the name, eighteen (the digits that appear to the right of the decimal point), is the numerator of the fraction. The second part of the name, thousandths, is the denominator of the fraction. 0.018 = $\frac{18}{1,000}$. The greatest common factor of 18 and 1,000 is 2, so we can reduce this fraction by dividing the numerator and denominator by 2: $\frac{18}{2}$ = 9, $\frac{1,000}{2}$ = 500; $\frac{18}{1,000}$ = $\frac{9}{500}$. For more on this concept, see Chapter 11.

48. To write a percent as a fraction, first write the percent as a decimal by moving the decimal point two places to the left: 29.41% = 0.2941. Now, write the decimal as a fraction. Read the name of the decimal aloud. 0.2941 is "two thousand, nine hundred forty-one ten thousandths." The first part of that name, *two thousand, nine hundred forty-one*, is the numerator of the fraction, and the second part of that name, *ten thousandths*, is the denominator of the fraction. 0.2941 = $\frac{2,941}{10,000}$. For more on this concept, see Chapter 11.

49. Write each number as a decimal so that they can be compared. To write a percent as a decimal, move the decimal point two places to the left. 81.57% = 0.8157. Write the fraction $\frac{31}{38}$ as a decimal so that it can be compared to 0.8157. Because the decimal 0.8157 has a digit in the ten thousandths place, we must round $\frac{31}{38}$ to the nearest hundred thousandth. 31 ÷ 38 = 0.8157894. . ., or 0.81579 to the nearest hundred thousandth. Add a trailing zero to 0.8157, line up the decimal points, and compare corresponding digits of the two numbers from left to right:

$$0.81570$$
$$0.81579$$

Both numbers have the same digits in the ones, tenths, hundredths, thousandths, and ten thousandths places, so we move to the hundred-thousandths place. The first number, 0.81570, has a 0 in the hundred thousandths place and the second number, 0.81579, has a 9 in the hundred thousandths place. Because $9 > 0$, $0.81579 > 0.81570$, and $\frac{31}{38} >$ 81.57%. For more on this concept, see Chapter 11.

50. Convert the fraction and the percent to decimals. The fraction can be converted to a decimal by dividing. $\frac{69}{96} = 69 \div 96 = 0.71875$. Compare this number to the given decimal, 0.7188:

$$0.71875$$
$$0.7188$$

Both numbers have the same digits in the ones, tenths, and hundredths places, but they have different digits in the thousandths place, so these numbers are not equal. Convert 71.88% to a decimal. Remove the percent sign and shift the decimal point two places to the left. 71.88% = 0.7188, which is equal to the given decimal. For more on this concept, see Chapter 11.

Glossary

Addend: The numbers being added in an addition sentence. In the number sentence 3 + 2 = 5, 3 and 2 are the addends.

Common denominator: A common denominator for two fractions is a number that is a multiple of each of the denominators of those fractions. For example, 15 is a common denominator for the fractions $\frac{2}{5}$ and $\frac{1}{3}$ because 15 is a multiple of 5 and a multiple of 3.

Complex fraction: A fraction whose numerator or denominator (or both) is a fraction. For example, $\frac{\frac{4}{5}}{3}$ is a complex fraction, as is $\frac{1}{\left(\frac{6}{7}\right)}$.

Decimal: A number that is written using one or more of ten symbols. The ten decimal symbols are 0, 1, 2, 3, 4, 5, 6, 7, 8, and 9.

Decimal point: A symbol that is used to separate the parts of a number that are 1 or greater from the parts of the number that are worth between 1 and 0.

Denominator: The bottom number of a fraction. In the fraction $\frac{2}{8}$, the denominator is 8. Because a fraction is a division statement, the denominator is the divisor of that statement.

Difference: The result of subtracting one number from another. In the number sentence 5 − 3 = 2, 2 is the difference.

Dividend: The number being divided in a division problem (the numerator of a fraction). For example, in the number sentence $4.5 \div 9 = 0.5$, 4.5 is the dividend.

Divisor: The number by which the dividend is divided in a division problem (the denominator of a fraction). For example, in the number sentence $4.5 \div 9 = 0.5$, 9 is the divisor.

"Equals" sign: =, used when two quantities are equal in value. For example, "the sum of 2 and 2 is equal to 4" can be represented with "$2 + 2 = 4$."

Equivalent fractions: Two or more fractions that have the same value. The fractions $\frac{1}{3}$, $\frac{2}{6}$, and $\frac{3}{9}$ are all equivalent to each other.

Factor: Each value being multiplied in a multiplication problem. For example, in the number sentence $1.2 \times 6 = 7.2$, 1.2 and 6 are factors.

Fraction: A number that represents a part of a whole. A fraction itself is a division statement. $\frac{2}{8}$ is an example of a fraction. In this fraction, 2 is divided by 8.

"Greater than" sign: >, used when the first of two values is larger than the second of those two values. For example, "5 is greater than 4" can be represented with "$5 > 4$."

Greatest common factor: The largest number that divides evenly into two numbers. For example, the greatest common factor of 12 and 18 is 6. Both 12 and 18 can be divided evenly by 6. Other numbers (1, 2, and 3) are also factors of both 12 and 18, but 6 is the greatest common factor.

Hundreds place: The third place to the left of the decimal point. In the number 123, 1 is in the hundreds place.

Hundredths place: The second place to the right of the decimal point. In the number 0.456, 5 is in the hundredths place.

Improper fraction: A fraction that has a value greater than or equal to 1, or, less than or equal to −1. These fractions have a numerator that is greater than or equal to the denominator. $\frac{3}{2}$ and $-\frac{3}{2}$ are both improper fractions.

Irrational number: A number that cannot be written in the form $\frac{x}{y}$ where x and y are integers and y is not equal to zero.

Leading zero: A zero that comes at the front of a number and does not affect the value of the number. The number 045 has a leading zero.

Least common multiple: A common multiple for two numbers is a value that is divisible by both numbers. The least common multiple for those two numbers is the smallest value that is divisible by both numbers. For example, 30 is a common multiple of 3 and 5, because 30 is divisible by both 3 and 5, but 15 is the least common multiple of 3 and 5, because 15 is divisible by both numbers, and there is no smaller value that is divisible by both 3 and 5.

"Less than" sign: <, used when the first of two values is smaller than the second of those two values. For example, "4 is less than 5" can be represented with "4 < 5."

Like fractions: Two or more fractions with the same denominator. For example, $\frac{1}{3}$ and $\frac{2}{3}$ are like fractions, because they both have a denominator of 3.

Minuend: In a subtraction sentence, the number from which you are subtracting. In the number sentence 5 − 3 = 2, 5 is the minuend.

Mixed number: A number that has a whole number part and a fraction part. The number $1\frac{1}{2}$ is a mixed number. Mixed numbers, like improper fractions, have a value that is greater than or equal to 1, or less than or equal to −1. The improper fraction $\frac{3}{2}$ is equal to the mixed number $1\frac{1}{2}$.

Non-terminating decimal: A decimal whose digits continue forever. A non-terminating decimal may repeat, such as 0.3333. . . If a non-terminating decimal does not repeat, then it is an irrational number. All rational numbers can be written as either terminating decimals or repeating decimals.

Numerator: The top number of a fraction. In the fraction $\frac{2}{8}$, the numerator is 2. Because a fraction is a division statement, the numerator is the dividend of that statement.

Ones place: The first place to the left of the decimal point. In the number 123, 3 is in the ones place.

Operand: Each number on which an operation (such as addition or subtraction) is performed. In the number sentence $5 - 3 = 2$, 5 and 3 are operands.

Percent: A ratio that represents a part-to-a-whole as a number out of one hundred. For example, $0.25 = \frac{25}{100} = 25\%$.

Percent decrease: The difference between an original value and a new value divided by the original value. The new value is subtracted from the original value, because the original value is larger than the new value. Percent decrease is used to show the decline from an original value to a new value.

Percent increase: The difference between an original value and a new value divided by the original value. The original value is subtracted from the new value, because the new value is larger than the original value. Percent increase is used to show the growth from an original value to a new value.

Place value system: A system that uses a number system, such as the decimal system, and the position (or place) of each digit in a number to determine what value each digit in the number is worth.

Product: The result of multiplication. For example, in the number sentence $1.2 \times 6 = 7.2$, 7.2 is the product.

Proper fraction: A fraction that has a value less than 1 and greater than -1. For example, $\frac{2}{3}$ is a proper fraction, as is $-\frac{2}{3}$.

Proportion: A relationship that shows two equivalent fractions, or ratios. For example, the fractions $\frac{3}{4}$ and $\frac{6}{8}$ are in proportion to each other because $\frac{3}{4} = \frac{6}{8}$.

Quotient: The result of a division problem. For example, in the number sentence $4.5 \div 9 = 0.5$, 0.5 is the quotient.

Ratio: A comparison, or relationship, between two numbers. Ratios are often shown with a colon, but they can also be expressed as fractions. The ratio "2 to 1" can be written as 2:1 or $\frac{2}{1}$.

Rational number: A number that can be written in the form $\frac{x}{y}$, where x and y are integers and y is not equal to zero.

Reciprocal: The reciprocal of a fraction is the multiplicative inverse of the fraction. To find the reciprocal of a fraction, switch the numerator and the denominator. The reciprocal of $\frac{3}{4}$ is $\frac{4}{3}$.

Repeating decimal: A decimal in which the same digit or sequence of digits repeat over and over, infinitely. A single digit could repeat, such as 0.33333. . ., or a sequence of digits might repeat, such as 0.297297297. . . . The repetition can begin before or after the decimal point, and not necessarily at the first digit after the decimal point (the tenths digit). For example, the decimal 0.916666666. . . repeats starting with the thousandths digit. A repeating decimal is an example of a non-terminating decimal.

Rounding: The process of taking a number and making it less precise by removing one or more digits from the end of the number, replacing those digits with zeros if necessary.

Subtrahend: In a subtraction sentence, the number that you are subtracting. In the number sentence 5 – 3 = 2, 3 is the subtrahend.

Sum: The result of adding two numbers. In the number sentence 3 + 2 = 5, 5 is the sum.

Tens place: The second place to the left of the decimal point. In the number 123, 2 is in the tens place.

Tenths place: The first place to the right of the decimal point. In the number 0.456, 4 is in the tenths place.

Terminating decimal: A decimal that does not continue infinitely. In other words, a terminating decimal is a value that can be represented fully with a finite number of digits. For example, 0.25 is a terminating decimal.

Thousandths place: The third place to the right of the decimal point. In the number 0.456, 6 is in the thousandths place.

Trailing zero: A zero that comes at the end of a number and does not affect the value of the number. The number 45.0 has a trailing zero.

Unlike fractions: Two or more fractions with different denominators. For example, $\frac{1}{3}$ and $\frac{1}{4}$ are like unlike fractions, because they do not have the same denominator.

Whole number: A positive number that has no fractional part—no digits to the right of the decimal point. For example, 45 is a whole number.

Notes

Notes